VETERINARY ZOOTOXICOLOGY

Murray E. Fowler, D.V.M.
Diplomate
American Board of Veterinary Toxicology
American College of Zoological Medicine
American College of Veterinary Internal Medicine
Professor Emeritus
Department of Medicine
School of Veterinary Medicine
University of California
Davis, California

CRC Press
Taylor & Francis Group
Boca Raton London New York

CRC Press is an imprint of the
Taylor & Francis Group, an informa business

First published 1993 by CRC Press
Taylor & Francis Group
6000 Broken Sound Parkway NW, Suite 300
Boca Raton, FL 33487-2742

Reissued 2018 by CRC Press

Library of Congress Cataloging-in-Publication Data

Fowler, Murray E.
Veterinary zootoxicology / Murray E. Fowler.
p. cm.
Includes bibliographical references and index.
ISBN 0-8493-6791-3
1. Veterinary toxicology. 2. Poisonous animals—Toxicology.
I. Title.
SF757.5.F69 1992
636.089'594—dc20 92-30156

ISBN 13: 978-1-315-89843-8 (hbk)
ISBN 13: 978-1-351-07753-8 (ebk)

Visit the Taylor & Francis Web site at http://www.taylorandfrancis.com and the
CRC Press Web site at http://www.crcpress.com

PREFACE

Snakebite and the bites and stings of other venomous animals have always captured the imagination of the public. Physicians have diagnosed and treated venomous bites and stings since antiquity. Likewise, veterinarians must deal with bites and stings with the added difficulty that the animal that inflicted the bite or sting usually cannot be identified. No single source of information on poisonous and venomous animals and their effects on domestic and wild animals is currently available to the veterinarian.

Some of the veterinary literature is based on data extrapolated from human envenomation reports. Such extrapolation is not always valid. Much misinformation and legend are associated with writings about venomous animals.

The last decade has been the age of enlightenment as the techniques of modern technologies of biochemistry, physiology, pharmacology, toxinology, and toxicology have begun to be applied to the study of complex venoms and their constituents. Investigators are beginning to unravel the mysteries of venoms. Why are some venoms so potent? What is the precise chemical structure and how does it relate to the pharmacologic effect of the venom? Many of the toxins in venoms are enzymes that catalyze fundamental reactions in metabolic pathways. A knowledge of these fundamental factors provides the framework for a more rational approach to first aid and therapy.

The author is a clinician, with a strong interest in clinical toxicology relating to domestic and wild animals. The objective of this book is to provide a general overview of the state of the art in venom research, with an aim to provide background for the rational treatment and management of poisoning and envenomation.

The biochemist or toxinologist will find the volume ludicrously superficial, but it is hoped that the practicing veterinarian, student, teacher, wildlife biologist, and others who must deal with poisoning and envenomation of animals on a clinical basis may find the tools to evaluate, diagnose, treat, and manage suspected or known cases of animal envenomation.

The literature on poisonous and venomous animals has been perused to expand upon the author's knowledge and experience. The emphasis is on clinical problems encountered in the U.S., but a world perspective is included, with the stipulation that it is the author's interpretation of the available world literature.

The ultimate goal of the book is to provide facts, techniques, methodologies, and regimens for more effective clinical management of animals poisoned or envenomated by other animals.

Murray E. Fowler, D.V.M.

This volume could not have been produced without the special journalistic skills of my wife, Audrey C. Fowler. She not only provided moral support, but copyread and proofed the entire manuscript, did a share of the typing, and provided advice on all aspects of the project.

This book is dedicated, with love, to Audrey C. Fowler.

TABLE OF CONTENTS

VETERINARY ZOOTOXICOLOGY

PART I: Introduction

Chapter 1

AN OVERVIEW OF POISONOUS AND VENOMOUS ANIMALS

I. INTRODUCTION

Venomous and poisonous animals are found in most of the animal phyla (Table 1). This chapter is not meant to be a definitive discourse on animal taxonomy. In fact, this is impossible because even taxonomists do not agree on all aspects of animal classification. Many of the classical subgroupings are not included in Table 1 (e.g., subphylum, superfamily, and infraorder); rather, an attempt has been made to show the general relationships of the higher taxa. Additional details will be discussed in appropriate sections.

Approximately 45,000 species of vertebrate animals and millions of species of invertebrates live in habitats suitable for their way of life, interrelating with other animals, plants, and geologic formations. Most species of animals are utilized for food by predators. The predator/prey relationship is delicate, and one function of poisons and venoms is to dissuade attack by predators or other potential enemies. Another function is to obtain food.

II. BRIEF OVERVIEW OF THE ANIMAL KINGDOM

Representative poisonous species of animals may be found in all classes except *Aves*, from the simplest unicellular organisms (protozoa) to species of the higher orders of mammals. Many animal groups are strictly marine, not likely to poison terrestrial animals. Only minimal coverage of these groups will be given.

Chapters will be devoted to a discussion of the venomous species in the various classes and orders.

TABLE 1
An Overview of the Venomous and Poisonous Members of the Animal Kingdom

Kingdom — *Animal*
Phylum — *Protozoa*
  Class — *Mastigophora* — *flagellates*
    Order — *Dinoflagellata*
      Family — *Peridiniidae*
      Gonyaulax cantenella
      Gonyaulax tamarensis
Phylum — *Porifera* — sponges
      Microciona prolifera — red sponge

TABLE 1 (continued)
An Overview of the Venomous and Poisonous Members of the Animal Kingdom

- Phylum — *Coelenterata (Cnidaria)* — anemone, jellyfish
  - Class — *Hydrozoa*
    - Order — *Siphonophora*
      - *Physalia physalis* — Portuguese man-of-war
  - Class — *Scyphozoa* — jellyfish, sea wasps
  - Class — *Anthozoa* — anemones, corals
- Phylum — *Platyhelminthes* — tapeworms, flukes
- Phylum — *Nematoda* — round worms
- Phylum — *Mollusca* — shellfish, cones
  - Class — *Cephalopoda* — octopus — 650 spp.
  - Class — *Gastropoda* — snails, seashells — 33,000 spp.
      - Family — *Conidae* — cones — 400–500 spp.
- Phylum — *Annelida* — segmented worms
      - *Lumbricus terrestris* — earthworm, night crawler
- Phylum — *Arthropoda* — joint-legged animals
  - Class — *Arachnida* — spiders, scorpions, ticks, mites
    - Order — *Acarina* — ticks, mites
    - Order — *Araneida (Araneae)* — spiders
    - Order — *Scorpionida* — scorpions
  - Class — *Crustacea* — shrimp, copepods
  - Class — *Diplopoda* — millipedes
  - Class — *Chilopoda* — centipedes
  - Class — *Insecta* — insects
    - Order — *Hemiptera* — triatomes
    - Order — *Coleoptera* — blister beetles
    - Order — *Lepidoptera* — butterflies, moths
    - Order — *Diptera* — flies, mosquitos
    - Order — *Hymenoptera* — ants, bees, wasps
- Phylum — *Echinodermata* — starfish, urchins
      - *Acanthaster planci* — crown-of-thorns starfish
- Phylum — *Chordata* — vertebrate animals
  - Class — *Elasmobranchii* — sharks, rays
  - Class — *Osteichthyes* — bony fishes
  - Class — *Amphibia* — amphibians
    - Order — *Urodela (Caudata)* — salamanders, newts
    - Order — *Salientia (Anura)* — frogs and toads
  - Class — *Reptilia* — reptiles
    - Order — *Squamata* — snakes and lizards
    - Order — *Testudines (Chelonia)* — turtles, tortoises — 400 spp.
    - Order — *Crocodilia (Loricata)* — alligators, crocodiles — 25 spp.
  - Class — *Aves* — birds, no venomous or poisonous species
  - Class — *Mammalia* — mammals
    - Order — *Monotremata* — platypus, echidna
      - *Ornithorhynchus anatinus* — duck-billed platypus
    - Order — *Insectivora* — shrews, solenodons
    - Order — *Carnivora* — polar bear
    - Order — *Pinnipedia* — seals, sea lions, walrus

REFERENCES

Halstead, B. W., *Poisonous and Venomous Marine Animals of the World,* Revised ed., Darwin Press, Princeton, NJ, 1978.

Grzimek, H. C., Ed., *Grzimek's Animal Life Encyclopedia,* Vol. 1–13, Van Nostrand Reinhold, New York, 1975.

Caras, R., *Venomous Animals of the World,* Prentice-Hall, Englewood Cliffs, NJ, 1974.

Storer, T. E., Usinger, R. L., Stebbins, R. C., and Nybakken, J. W., *General Zoology,* 6th ed., McGraw-Hill, New York, 1979.

Hickman, C. P., Sr., Hickman, C. P., Jr., and Hickman, F. M., *Integrated Principles of Zoology,* 5th ed, C. V. Mosby, Saint Louis, MO, 1974.

Chapter 2

A HISTORY OF ZOOTOXICOLOGY

I. INTRODUCTION

A poison is any substance which, when ingested, inhaled, absorbed, or when applied to, injected into, or developed within the body, in relatively small amounts, may cause damage to body structure or disturbance of function through its chemical action. Poisons and poisonous animals have been of interest to humans since the dawn of creation. They played a part in the everyday lives of primitive people that was recorded by them in the earliest writings. The science of toxicology (the study of poisons) is of more modern derivation, along with that of biotoxicology (the study of poisons from living organisms) and zootoxicology (the study of poisonous animals).

It is the latter subject that will be addressed in this book. However, an overview of the history of toxicology may aid in obtaining a perspective of this rapidly expanding discipline. Advances in the fields of analytic chemistry, biochemistry, and molecular biotechnology are now making it possible to isolate and purify individual components, using minute amounts of a venom or toxin. The toxinologist can study a single fraction (venin) within a venom to delineate its pharmacologic action and contribution to the toxicity of the crude venom.

Any field of science has a language that must be understood to communicate information. A glossary is provided at the conclusion of this chapter, defining a few of the words that will be used frequently throughout the text (Table 1). Table 2 also provides foreign language equivalents to English words for those who may wish to peruse foreign references.

This book is not intended to be a definitive review of the entire field of zootoxicosis, but rather an overview, with emphasis on the clinical diagnosis and management of animal envenomation.

II. TOXICOLOGY IN ANTIQUITY

Medicine is an ancient science. The use of plant and animal substances to heal disease is as old as the diseases themselves. The pharmacologic effects of plant and animal substances were derived from the presence of plant secondary metabolites (compound)(PSM) or venoms and poisonous substances from animals. It is now known that PSMs provide the plant or animal with the ability to survive in a special niche to avoid predation, defend against enemies, or obtain food. The determining factor of whether a substance is a medicine or a poison may be a matter of dosage. Early practitioners of medicine were priests or magician/physicians, and it was commonly thought that only those guilty of a crime would be affected by a poison.

A synopsis of the history of zootoxicology follows. Readers who wish more detailed historical information are referred to Halstead[1] and the references that he cites.

A. MESOPOTAMIA (BABYLON AND ASSYRIA)

The development of human civilization is thought to have originated in the fertile crescent which is now occupied by Iraq, Syria, Israel, and Egypt, but was known as Babylon and Assyria historically. Medicine had its beginnings in this region. The code of Hammurapi was chiseled into stone, providing evidence for future generations of the expected behavior of physicians.

Bites and stings from venomous animals occurred and were treated with scarification, bloodletting (cut and suction), the use of incantations, and a procedure called cupping (use of an animal horn to neutralize the poison).[1]

B. EGYPT

The Egyptian civilization developed at approximately the same time as those of Babylon and Assyria. Writing paper was manufactured from the papyrus plant, *Cyperus papyrus* — sedge family — *Cyperaceae,* which is a reed-like plant that grows in shallow water. Modern humans have learned of the prescriptions and medicine used for treating ailments through various papyri. The Smith papyrus (1600 B.C.) lists charms for use against snakebite, and the Hearst Medical Papyrus provides numerous prescriptions, some of which were prescribed for animal bites.[1] One Egyptian picture story depicts a puffer fish, *Tetraodon* sp. which was recognized as a poisonous fish.

C. ISRAEL

Ancient Hebrew people were guided in every aspect of their lives by the Old Testament. Strict dietary laws were imposed on the people, presumably to maintain health and prevent disease (e.g., trichinosis in pork). Early Israelites were admonished not to eat items from the water that had neither fins nor scales (Deuteronomy 14:9–10). Fish with neither of these structures tend to be the most poisonous and thus should be avoided.

One of the plagues pronounced on the Egyptians by Moses was the earliest account of red tide caused by dinoflagellates (the water turned to blood, killing all the fish — Exodus 7:19–20).

D. CHINA

Ancient and modern Chinese medicine rely heavily on natural products for healing disease (herbal medicine, rhinoceros horn, and gall bladders). Medical practitioners recognized scaleless fish as being toxic and the first report of ciguatera toxicosis was described from ingestion of yellowtail tuna (ca A.D. 700).[1]

E. GREECE

Modern medicine had its genesis in the Greek civilization, from 500 B.C to A.D. 500. Medicine became separated from myth, magic, and religion.

Hippocrates, the Father of Medicine, knew of the toxicity of sea urchin spines. Aristotle (384 to 322 B.C.) knew of the stinging of jellyfish. Alexander the Great forbade his troops to eat fish, presumably because of concern for toxicity.

The king of Pontus, Mithridates VI, (120 to 63 B.C.) was one of the first to advance the art and science of poisons and antidotes. He experimented to attempt development of a universal antidote, including numerous substances in his concoctions, which were called "mithridates". Even today, the term *mithridates* or "shotgun treatment" is used to describe a situation in which multiple drugs are prescribed, in the hope that one of the drugs will be effective against the ailment suffered by the patient. Mithridates' philosophy was followed for many centuries in treating all kinds of diseases, including poisoning and envenomation.

F. ROME

Roman civilization began during the Greek era and ultimately supplanted it. Discordes (A.D. 50 to 100) wrote the first *materia medica;* it served the medical profession for 1600 years. He describes treatment for various marine animal intoxications, including the stingray and "sea vipers".

Pliny the Elder (A.D. 29 to 79) wrote the classic *Naturalis Historia,* a comprehensive encyclopedia of the natural world, which was the natural history "bible" until the Renaissance. Some of the earliest written records of the poisonous or venomous qualities of animals may be found in this volume. Pliny wrote one section devoted to *De Venenalis Marinus* (marine poisons).

G. ISLAMIC COUNTRIES

Moslem countries provided little original contribution to the field of toxicology until a Persian by the name of Geber ibn Hajan (ca. A.D. 750 to 760) wrote a long treatise on poisons, including venoms and their antidotes.

III. MEDIEVAL PERIOD

The collapse of the Roman empire ushered in a period of intellectual stagnation (Dark Ages). Many of us have the impression that nothing was accomplished in the sciences during this time; but in reality, the printing press and gun powder, among other things, were invented during the Dark Ages. A few physicians wrote about venomous animals, but little original toxicologic thought arose in this era.

IV. MODERN PERIOD

With the Renaissance came the age of scientific inquiry, and numerous naturalists, physicians, missionaries, explorers, and historians began to document the prevalence of intoxication of humans by animals and plants. Little was known about the nature of the venoms or poisons. The provincialism of European and Mediterranean country writers was overcome as explorers returned from far-flung expeditions with factual and fanciful tales of the tropical regions of the world. Captain James Cook of H.M.S. *Resolution* nearly died of tetrodotoxin poisoning after consuming fish in New Caledonia.

Field and laboratory investigations were conducted to verify anecdotal reports of poisoning and envenomation. It was readily accepted that some plants and animals were poisonous to animals, but reasons why the poisons were there in the first place were less well understood.

Carl von Linne (Linnaeus), the Swedish naturalist, made a monumental contribution to biology and toxicology when he introduced the system of binomial naming and classification of plants and animals *(Systema Naturae*, 1758). His system enabled scientists from around the world to communicate about plants and animals without having to know all the local, common names.

Scientific disciplines began to proliferate (chemistry, physiology, pharmacology, and toxicology). Scientific meetings provided a forum for expounding theories, and professional publications spread information throughout the world. Biologists studied the natural history of plants and animals. Anatomists dissected and described the venom apparatus. Biochemists studied venoms and poisons, and much was learned in the 19th century about chemical structure and pharmacologic effects. Textbooks on toxicology were published in German, French, Japanese, English, and other languages. Toxicology became a required course in most medical and veterinary schools.

The field of toxicology became so broad that scientists began to narrow their fields of interest further: into plant poisoning, marine biotoxicology, or pesticide toxicology. Scientists of the 20th century continued and intensified the study of poisons. Halstead[1] provides a listing of marine toxicologist pioneers and current investigators (as of 1978).

A. 1950–1991

A noteworthy achievement of the latter half of the 20th century is the development of a better understanding of the ecology of poisons and venoms. Some answers have been discovered to such puzzling questions as: Why are plants and animals poisonous or venomous? What have poisons contributed to evolution? How can an understanding of poisons and venoms contribute to the management of both free-ranging wild animal populations and domestic animals? The following chapters will discuss these questions.

V. PLANT TOXIN ECOLOGY

Prior to 1950, toxicologists working with plants explained the presence of poisonous substances in plants as an aberration of nature, a storage form of chemicals to be used later by the plant in metabolism or as an end-product of plant metabolism. Brower's historical report demonstrated conclusively that the presence of PSMs was, in fact, a defense against herbivory.[2] The concept that plants contain compounds that serve as defenses against herbivory was not new. Stahl[3] suggested this theory in 1888, but proof was not demonstrated until the sophisticated biochemical technology of the early 1950s provided the tools for solving the riddles.

Other investigators soon found numerous instances where insects abstained from consuming certain plants because of the unpalatable or toxic chemicals contained therein. On the other hand, certain insects developed methods for coping with the toxin or actually utilizing it to their own advantage (Monarch butterfly larvae, see page 87). The last 3 decades have witnessed a plethora of publications on this subject. A subdiscipline of biology (chemical ecology) has developed, based on the relationship of plants and animals: the toxins produced by the plant and the detoxification mechanisms of the animal.[2]

It is now understood that plants and herbivorous animals (primarily insects) coevolved in an adversarial relationship. Plant populations developed various strategies for coping with herbivores that could potentially destroy them. Mechanical means included thorns, spines, and harsh outer coats of stems and leaves.[4-6] More important are the chemical deterrents in the form of PSMs. These chemicals are either bitter tasting, offensively odorous, poisonous, or have antinutritional effects.[7] As plant populations became more efficient in the production of PSMs or developed a new PSM, animals that normally fed on the plants likewise improved their own methods of coping.

The case for coevolution of plants and insects is easily demonstrated, while evidence for plant/mammal coevolution is more elusive. This seems logical, for insects and plants have shared the earth for approximately 250 million years (Carboniferous Period, Paleozoic Era), while primitive mammals have been on earth only for 100 million years (Cretaceous Period, Mesozoic Era). The progenitors of large, wild herbivorous animals such as cattle, horses, bison, and antelope didn't appear until 40 million years ago, and all the domestic animals arose from their wild counterparts less than 12,000 years ago.

Even though mammalian herbivores may not have coevolved with plants like insects, there is sound evidence that plant (dietary) selection is determined by the presence and quantity of PSMs in the part of the plant consumed (leaves, stems, fruit, and seeds).[8]

VI. ANIMAL TOXIN ECOLOGY

Animal toxins and venoms are used to obtain food, deter predation, regulate population, aid in exploiting a territory, and aid in defense against enemies. Just as in plants, the end products of secondary metabolism are channeled to storage sites within or upon the animal's body, where there is slow turnover. The slow turnover minimizes the necessity of excessive expenditure of precious energy and chemical resources to this function.

VII. CONCLUSIONS

Poisons and venoms have served a vital role in the evolution of plants and animals. Scientists are just beginning to unravel many of the mysteries of plant/animal/animal interactions.

TABLE 1
Glossary

Anaphylaxis — A manifestation of immediate hypersensitivity in which exposure of a sensitized animal to a specific antigen results in dyspnea, followed by vascular collapse, and shock accompanied by urticaria, pruritus, angioedema, and pulmonary edema.
Anticoagulant factor — A toxin found in some venoms that inhibits blood coagulation and promotes hemorrhage.
Antivenin — A product composed of antibodies to specific venins and produced by hyperimmunizing horses with the venin or its parent substance, the venom.
Antivenom — Sometimes used interchangeably with antivenin, but is not the preferred term in the U.S.
Biotoxin — Any poisonous substance produced by and derived from a living organism, either plant or animal.
Coagulant factor — A toxin found in some venoms that stimulates intravascular blood clotting, with the subsequent risk of thrombus formation and disseminated intravascular clotting (DIC).
Envenomation — The act of injecting a venom into the tissue of another animal. Sometimes referred to as *envenoming* in the literature.
Hematoxin (hematotoxin) — Poisonous to the blood and hematopoietic system.
Hyaluronidase activity — Enzyme(s) that catalyze the breakdown of hyaluronic acid, which serves as a bonding agent between cells. The breakdown allows more rapid spread of the venom.
Intoxication — Poisoning, the state of being poisoned.
Necrotoxin (cytotoxin) — A toxin that kills tissue cells.
Neurotoxin — A toxin that is poisonous to or destroys nerve tissue.
Poison — Any substance which, when ingested, inhaled, or absorbed, or when applied to, injected into, or developed within the body, in relatively small amounts, by its chemical action may cause damage to structure or disturbance of function (toxin).
Poisoning — The morbid condition produced by a poison (intoxication).
Poisonous — Pertaining to, due to, or of the nature of a poison (venomous).
Serum sickness — A complicated immune response confined to the development of circulating immune complexes following injection of an antigen (e.g., snake antivenin or equine serum). Halliwell: The complexes become deposited in the basement membranes of synovial membranes, kidneys, or vascular endothelium, causing vasculitis, arthritis, or glomerulonephritis. The reaction occurs days to weeks following the use of the antivenin. Serum sickness is more common in humans than in animals and is less common now with more sophisticated technologies for purification of immunoglobulins.
Sting — An injury caused by the venom of an animal and introduced into another animal by a specialized apparatus called a "stinger".
Toxic — Pertaining to, due to, or of the nature of a poison or toxin.
Toxicant — A poisonous agent.
Toxication — Poisoning.
Toxicity — The quality of being poisonous, especially the degree of virulence of a poison; usually expressed as the quantity of dried venom that kills 50% of a test animal (LD_{50}) in milligrams of dried venom per kilogram of body weight (mg/kg). More precisely, this should be called *lethality*, which is only one of the toxic manifestations of envenomation. In animal envenomation, cytotoxic or necrotoxic effects may be more important.
Toxicosis — The disease caused by a toxin or poison.
Toxin — A poison. As used in this text, it refers to a specific chemical substance present in a venom (venin), and not to the proteinaceous substance produced by bacteria or other microorganisms.
Toxinology — The study of toxins.
Toxinologist — A person who studies toxins.

TABLE 1 (continued)
Glossary

Venenate — Synonymous with envenomate or envenomation.
Venenation — The same as venenate.
Venin — One of the specific toxins found in a venom.
Venom — A poison; specifically, a toxic substance normally produced by an animal which, when injected into another animal, produces a poisonous effect. Venoms have a complex composition.
Venomation — The same as envenomation (preferred).
Venomous — Secreting a venom (poisonous).
Zootoxin — A toxic substance of animal origin, such as venomous snakes, spiders, and scorpions.
Zootoxicosis — The disease caused by an animal venom or toxin.

From *Dorland's Illustrated Medical Dictionary,* 27th ed., W. B. Saunders, Philadelphia, 1985. With permission.

TABLE 2
Foreign Equivalents for English Words Used in Zootoxicology

English	German	French	Spanish	Portuguese	Italian	Latin	Greek
Antivenin	Gegengift	Antivenin	Contraveneno, antídoto	Contraveneno, antiveneno	Antiveleno		
Biotoxin	Biotoxin	Biotoxine	Biotoxina	Biotoxina	Tossina biologica		
Envenomation	Vergiften	Envenimation; Envenimer (verb)	Envenenar, envenenamiento	Envenenar	Avvelenamento da morsi		
Intoxication	Intoxikation	Intoxication	Intoxicación	Intoxicação	Intossicazione		
Poison	Gift	Poison, venin, toxique	Veneno	Veneno	Veleno	Venenum, toxicum	Toxikon
Poisoning	Vergiftung	Empoisonnemement	Envenenamiento, venenación	Envenenamento	Avvelenamento	Veneficum	
Poisonous	Giftig	Toxique, venimeux	Venenoso	Venenoso	Velenoso		
Sting	Stechen, to sting; der Stich, the	Dard, piqure (noun), piquer (verb)	Aquijón, picadura	Aguilhao, ferrao, picada, ferroada	Pungiglione		
Stinger	sting; der Stachel, the apparatus						
Toxic	Giftwirkung, toxisc	Toxique	Tóxico	Tóxico	Tossico		
Toxicity	Giftigkeit, toxizität	Toxicité	Toxicidad	Toxicidade	Tossicita		
Toxin	Toxin, gift giftstoff	Toxine	Toxina	Toxina	Tossina, sostanza tossica		Toxikon
Venin	Venin	Venin	Venin				
Venom	Tiergift	Venin	Veneno	Veneno	Veleno de serpente	Venenum	Toxikon
Venomous	Giftig	Venimeux	Venenoso	Venenoso	Velenoso	Venenatus	
Zootoxin	Tiergift	Zootoxine, toxine animale	Toxina animal	Toxina animal	Tossina animale		

REFERENCES

1. **Halstead, B. W.,** *Poisonous and Venomous Marine Animals of the World,* Revised ed., Darwin Press, Princeton, NJ, 1978, 1.
2. **Brower, L. P.,** Ecological chemistry, *Sci. Am.,* 220, 22, 1969.
3. **Stahl, E.,** Planzen und Schnecken. Eine biologische Studie uber die Schultzmittal der Planzen gegen Schneckenfrass, *Jena Zentralblatt Med. Nat.,* 22, 557, 1888.
4. **Fowler, M. E.,** Plant poisoning in free-living wild animals: a review, *J. Wildlife Dis.,* 19, 34, 1983.
5. **Laycock, W. A.,** Coevolution of poisonous plants and large herbivores on rangelands, *J. Range Manage.,* 31, 335, 1978.
6. **Freeland, W. J. and Janzen, D. H.,** Strategies in herbivory by mammals: the role of plant secondary compounds, *Am. Nat.,* 108, 269, 1974.
7. **Harborne, J. B.,** The chemical basis of plant defense, in *Plant Defenses Against Mammalian Herbivory,* Palo, R. T. and Robbins, C. T., Eds., CRC Press, Boca Raton, FL, 1991, 45.
8. **Freeland, W. J.,** Plant secondary metabolites: Biochemical coevolution with herbivores, in *Plant Defenses Against Mammalian Herbivory,* Palo, R. T. and Robbins, C. T., Eds., CRC Press, Boca Raton, FL, 1991, 62.

PART II: Marine Toxins

Chapter 3

MARINE ZOOTOXINS

I. INTRODUCTION

Numerous marine organisms produce substances that are poisonous. Some are toxic only when ingested by other animals, while others possess an anatomical apparatus for injecting venom. Marine zootoxins are among the most highly toxic substances known (Table 1). Biologists have not unraveled all of the mysteries, but much is known about the role of toxins in predator/prey relationships, food gathering, and defense mechanisms.

Although livestock and pets are not often poisoned by marine organisms, they sometimes eat or encounter them. Free-ranging marine animals may be at risk of envenomation or poisoning.

II. PARALYTIC SHELLFISH POISONING

A. IDENTIFICATION AND BIOLOGY

Paralytic shellfish poisoning is caused by ingestion of shellfish (clams, mussels, scallops, and oysters) that have, in turn, ingested toxic dinoflagellates.[1]

1. Dinoflagellates

Dinoflagellates are marine plant protozoa and are major components of the phytoplankton that forms the food base for millions of marine organisms. Under seasonally rich environmental conditions, dinoflagellates proliferate exponentially, resulting in a "bloom". An intense bloom produces so many organisms that they color the water. Some common red protozoa give rise to what is referred to as "red tide". Dinoflagellate bloom depends on favorable weather conditions and an abundance of nutrient chemical substances in the water.[1] Species of dinoflagellates known to be toxic are listed in Table 1.

During red tides, massive mortality of fish and other marine organisms that consume phytoplankton has been noted. Freshwater algal blooms may also be poisonous to terrestrial animals.[2]

2. Shellfish

The most common species of shellfish involved in paralytic shellfish poisoning are listed in Table 2. Shellfish are resistant to dinoflagellate toxins. Whether a given species of shellfish becomes toxic depends on the feeding behavior of the shellfish, the concentration of toxic dinoflagellates in the water, and the duration of bloom.

TABLE 1
Relative Potency of Toxins Administered Intraperitoneally in Mice

Toxin	LD_{50} (μg/kg)
Ciguatoxin[a]	0.45
Saxitoxin[a]	3.0
Tetrodotoxin[a]	8.0
Botulinum A	0.000026
d-Tubocurare	200.0
Brevetoxin a[a]	95.0
Chlorinated biphenol (dioxin)	2.0

[a] Marine toxins.

TABLE 2
Classification of Poisonous and Venomous Marine Animals

I. Phylum — *Protozoa (Protista)* — unicellular organisms
- A. Class — *Mastigophora* — flagellates
  - 1. Order — *Dinoflagellata* — dinoflagellates
    - a. Family — *Peridiniidae*
      - *Gonyaulax (Protogonyaulax) cantenella*
      - *Gonyaulax (Protogonyaulax) tamarensis*
      - *Gymnodinium (Ptychodiscus) breve*
      - *Gambierdiscus toxicus*

II. Phylum — *Porifera* — sponges
- *Microciona prolifera* — red sponge

III. Phylum — *Coelenterata (Cnidaria)* — anemone, jellyfish
- A. Class — *Hydrozoa*
  - 1. Order — *Siphonophora*
    - *Physalia physalis* — Portuguese man-of-war
- B. Class — *Scyphozoa* — jellyfish, sea wasps
  - a. Family — *Chirodropidae*
    - *Chironex fleckeri* — sea wasp, box jelly
    - *Chiropsalmus quadrigatus* — sea wasp
- C. Class — *Anthozoa* — anemones, corals in the following genera
  - Anemones: *Actinia, Actinodendron, Adamsia, Alicia, Anemonia, Calliactis, Lebrunia, Physobrachia, Rhodactis, Sagartis,* and *Telmatactis*
  - Corals: *Acropora, Astreopora,* and *Goniopora*

IV Phylum — *Mollusca* — shellfish, cones
- A. Class — *Cephalopoda* — octopus, squid, cuttle fish
  - *Octopus maculosa* — blue-ringed octopus
- B. Class — *Gastropoda* — snails, seashells
  - a. Family — *Conidae* — cones
    - *Conus* spp. — piscivorous cones
- C. Class — *Bivalvia (Lamellibranchiata, Pelecypoda)*
  - 1. Order — *Heterodonta* — razor clams, mud clams
    - *Mya arenaria* — soft-shelled clam — paralytic shellfish poisoning
    - *Saxidomus giganteus* — Alaska butter clam — paralytic shellfish poisoning

TABLE 2 (continued)
Classification of Poisonous and Venomous Marine Animals

V. Phylum — *Annelida* — segmented worms
- *Lumbriconereis heteropoda*
- *Glycera convoluta* — blood worm

VI. Phylum — *Echinodermata* — starfish, urchins, sea cucumbers
- *Diadema* spp. — long-spined sea urchin

VII. Phylum — *Chordata* — vertebrate animals
- A. Class — *Elasmobranchii* — sharks and rays
- B. Class — *Osteichthyes* — bony fishes. Numerous species of fish are either poisonous or venomous. (See Table 3 for more detail.)
- C. Class — *Reptilia* — reptiles
  - 1. Order — *Squamata* — snakes, lizards
    Suborder — *Ophidia (Serpentes)* — snakes
    - a. Family — *Hydrophiidae* — sea snakes
      - *Enhydrina schistosa* — sea snake
      - *Pelamis platurus* — yellow-bellied sea snake
  - 2. Order — *Chelonia (Testudinata)* — turtles
    - *Eretmochelys imbricata* — hawksbill turtle
    - *Dermochelys coriacea* — leatherback turtle
    - *Chelonia mydas* — green turtle
- D. Class — *Mammalia* — mammals
  - 1. Order — *Cetacea* — whales, dolphins
    - *Balaenoptera borealis* — sei whale
    - *Delphinapterus leucas* — beluga whale
    - *Physeter catodon* — sperm whale
    - *Neophocaena phocaenoides* — Asiatic porpoise
  - 2. Order — *Carnivora* — carnivores
    - a. Family — *Ursidae* — bears
      - *Thalarctos maritimus* — polar bear
  - 3. Order — *Pinnipedia* — seals, sea lions, walrus
    - a. Family — *Odobenidae* — walrus
      - *Odobenus rosmarus* — walrus
    - b. Family — *Otariidae* — sea lions, fur seals
      - *Neophoca cinerea* — Australian sea lion
    - c. Family — *Phocidae* — seals
      - *Erignathus barbatus* — bearded seal
      - *Pusa hispida* — ringed seal

3. Distribution of Poisoning

Gonyaulax catenella inhabits the Pacific coast of North America. *Gonyaulax tamarensis* is found on the east coast of North America and *Gymnodinium breve* inhabits the Gulf of Mexico and Florida. Other species are found elsewhere in the world.

B. CONDITIONS OF POISONING

Most outbreaks of paralytic shellfish poisoning have been associated with blooms of dinoflagellates. However, the toxin also has been isolated from shellfish when toxic dinoflagellates are low in numbers or absent from an area.

Massive fish mortality may be associated with a red tide, but this is thought to arise from toxic dinoflagellate cell lysis and toxin liberation upon passage across gill filaments. There is no evidence to suggest that acutely poisoned fish, or fish caught during a red tide, accumulate toxin to levels that are a hazard to humans.[3] Domestic pets are at risk only if fed leftover shellfish.

C. POISON

The first paralytic shellfish toxin purified and structurally characterized was isolated from clams and designated "saxitoxin". Since then, a number of similar toxins with slightly differing chemical configurations have been isolated (e.g., neosaxitoxin and gonyautoxin I–VIII). In fact, several toxins isolated from freshwater blue-green algae are structurally similar to saxitoxin.[4] Other toxins associated with shellfish include the neurotoxic brevetoxins a and b, and a gastroenteric toxin resembling akadaic acid.[3]

D. CLINICAL SIGNS

Paralytic shellfish poisoning results in severe neurotoxicity in humans. The syndrome described for humans begins with paresthesia of the orofacial region, progressing rapidly to the digits. This is followed by dizziness, ataxia, loss of muscle tone, inability to breathe, and death.[3,5] Signs may be variable depending on the relative and absolute amounts of the various toxins present.

E. DIAGNOSIS

Consumption of shellfish and the clinical syndrome are the primary means of diagnosis. No diagnostic analytical test is available to determine the presence of the toxin, and no lesions are present at autopsy.

F. TREATMENT

There is no antidote. Supportive care, especially assisted respiration, is important.

G. PROGNOSIS

The prognosis is good if respiratory action is maintained.

III. VENOMOUS COELENTERATES

A. IDENTIFICATION AND BIOLOGY

Representative venomous species of this large invertebrate group are listed in Table 2. As animal poisoning is rare; a detailed discussion of this group is not warranted for this book.

The infamous Portuguese man-of-war, *Physalia physalis*, is not a jellyfish, but rather a siphonophore, composed of a colony suspended from a float (pneumophore). The man-of-war has a sinister reputation, but it is probable that true jellyfish species are responsible for most of the deaths attributed to the

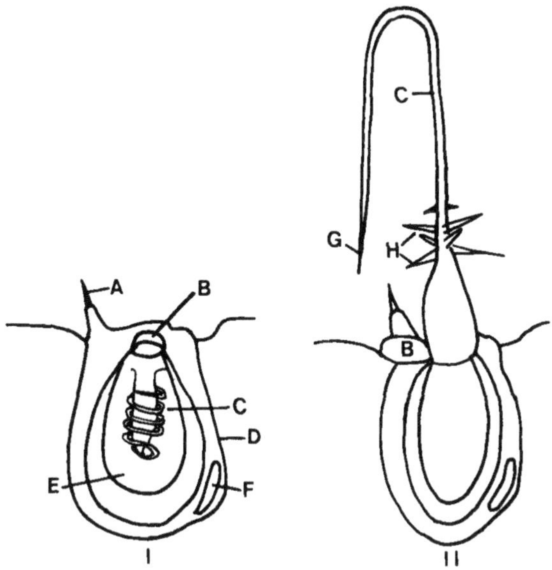

FIGURE 1. A diagram of quiescent nematocyst (I) and discharged nematocyst (II). A, cnidocil; B, operculum; C, thread tube; D, cnidoblast; E, venom capsule; F, nucleus; G, tip; and H, butt spines. (Adapted from Halstead and Vinci.[1])

man-of-war. Nonetheless, the man-of-war is a hazard to swimmers because the tentacles may trail for as far as 25 m.

Sea wasps (a true jellyfish) are the most dangerous species (see Table 2). Envenomation has been known to cause death in a matter of minutes.[6,7] Many other genera of jellyfish may inflict painful stings, but serious envenomation is unlikely.[6] Many sea anemones and corals have stinging capabilities (Table 2).

1. Venom Apparatus

Some coelenterates have long appendages (tentacles) that are covered with specialized cells (cnidoblasts), each containing a nematocyst (Figure 1).[7] The tentacles of anemones and corals are shorter, but also have nematocysts at the surface. The quiescent nematocyst contains a coiled thread tube and a fine hair (cnidocil) that acts as a trigger. When the cnidocil is disturbed, the nematocyst is everted, the tube uncoils and is thrust into the skin of the victim, and the venom expressed.

B. CONDITIONS OF POISONING

Coelenterates trail their tentacles through the water. Unwary prey species or larger animals come in contact with the tentacle and are envenomated. The primary use of the nematocyst and its venom is to obtain food, but the hunting method is indiscriminate. Human swimmers are stung if they come in contact with the tentacles of coelenterates. People may be also be envenomated if they touch jellyfish that have washed ashore. Thousands of jellyfish may be found on beaches after a storm, with nematocysts that may be triggered even after the death of the coelenterate.

Hairy mammals are less likely to be envenomated because the hair triggers the nematocyst before it touches the skin, except around the eyes, lips, nose, and foot pads.

C. VENOM

Although the nematocyst is the universal method of envenomation by this group of animals, venoms are by no means uniform throughout the group. Coelenterate venoms are complex mixtures of proteins, polypeptides, amines, enzymes, sterols, saponins, histamine, serotonin, and a variety of nonnitrogenous substances. Toxinologists are just beginning to isolate individual components and determine the contribution of each fraction to toxic manifestations.

D. CLINICAL SIGNS

1. Human

Man-of-war stings cause local and generalized pain. Once, as the author was scuba diving with friends off the coast of Mexico, a nondiver in the group decided to skin-dive in the vicinity of the dive boat. A short time later the skin-diver was found on the floor of the boat, writhing in excruciating pain. He had come in contact with man-of-war tentacles and immediately felt a burning sensation. Fortunately, he climbed at once into the boat, for within moments he was incapacitated from the pain and would have drowned had he remained in the water without assistance. Besides intense local pain, he suffered severe myalgia with periodic cramping. The victim was a veterinarian and he described the pain as pronounced spasms (like charley horses) of the muscles of the abdomen and limbs.

The dive site was many miles from the nearest physician, but fortunately meperidine (a narcotic analgesic) had been included in an emergency medical kit. Intravenous administration of meperidine abolished the pain and allowed the heart rate to return to normal. By the time the effects of the meperidine had dissipated, so had the systemic effects of the venom. Within 2 h, the victim resumed skin-diving.

Physicians treating *Physalia* victims have reported complaints of local pain, especially along the lymphatic drainage from the sting site, plus abdominal cramping and respiratory difficulties.[8] Erythema at the sting site may persist for days.[7] Corneal injury may also occur, persisting for weeks.

Stings from sea wasps, *Chironex fleckeri* and *Chiropsalmus quadrigiatus*, are much more severe and may be fatal to humans. The stings are extremely painful and may result in vesication and skin necrosis. Powerful, uncoordinated muscle spasms occur, causing dyspnea, prostration, and pulmonary edema.[7] Hypertension is an early sign, but later, blood pressure falls. Other cardiovascular effects include cardiac arrhythmias and collapse.[7]

Sea anemone stings cause mild, local pain with burning or itching as signs. Swelling and erythema may occur also. In more severe reactions, localized necrosis may occur and the ulcer may be slow to heal. Reported systemic signs include fever, chills, malaise, abdominal pain, nausea, vomiting, headaches, and prostration.[6]

Lacerations and frequently secondary infections, are the usual result if swimmers contact the calcarious base of corals, but some genera may envenomate also. Lesions and responses are similar to those of an anemone sting.

2. Animal

No reports of clinical poisoning in domestic animals have been published. However, in experimental studies, dogs, cats, rabbits, and laboratory rodents have shown responses similar to those of clinical envenomation in humans.

E. DIAGNOSIS

The pattern of erythema and swelling may aid in identification of some of the more important coelenterate stings in humans.[6] The hair coats of most domestic animal species preclude a linear pattern of response to tentacles.

F. TREATMENT

1. Human

Carefully remove any part of a tentacle that may remain on the victim, using sand, clothing, towel, or seaweed.[6] Application of suntan lotion or other skin creams may soothe the welts and help to remove any attached microscopic nematocysts. Topical application of cortisone ointments may relieve pain.

Therapy for systemic reactions is a medical procedure that may require respiratory assistance and symptomatic treatment for cardiovascular collapse or renal failure. An antivenin has been produced in Australia for sea wasp envenomation, but it is unlikely that antivenin will be available quickly enough to benefit the victim of a highly toxic sea wasp. Also in Australia, attempts are being made to immunize people who must engage in hazardous activities, against sea wasp venom.[7]

2. Animal

No references describe treatments for animals.

G. PROGNOSIS

Sea wasps are the only species that are likely to cause fatalities. The majority of fatal sea wasp stings have occurred in the waters near northern Australia, the Philippines, Malaysia, and in the Indian Ocean.[9] *Chironex fleckeri* has been identified in the Persian Gulf.*

*Personal communication, Dr. Richard Criddle, University of California, Davis, April 1992.

H. PREVENTION

Avoid stepping on or picking up jellyfish cast up on the seashore. Ask locals before swimming at unfamiliar beaches.

IV. POISONOUS AND VENOMOUS FISH

Marine fish contain numerous toxins or venoms. Categories of fish toxins are still changing as new toxins are isolated and identified. Table 3 provides an overview of known fish poisons and representative species commonly involved in poisoning. Numerous reviews, monographs, and current research papers discuss these matters in detail.[6,10-16]

V. CIGUATERA POISONING

A. BIOLOGY

Ciguatera poisoning is a circumtropical disease (or disease complex) caused by the consumption of certain tropical reef or in-shore fishes that accumulate toxins via the food chain.[16] Ciguatoxicosis was recorded in the Americas in the early 1500s, and European explorers of the tropics frequently reported fish poisoning.[16] Ciguatoxin is the primary toxin, but another, called "maitotoxin," is also important. Ciguatoxin is produced by the dinoflagellate *Gambierdiscus toxicus*.

The list of fish known to contain ciguatoxin may include 400 species, but the number of species containing the toxin in concentrations sufficient to cause poisoning is probably less than 20. A few of the more important fish implicated in poisoning are listed in Table 3. Typically, ciguatoxic fish are restricted to species feeding on algae or detritus around tropical reefs, or the larger carnivorous reef fish that prey upon herbivores.

B. CONDITIONS OF POISONING

Ciguatoxicosis occurs sporadically and unpredictably in tropical and subtropical, coastal continental, and insular regions. It is estimated that 10,000 to 50,000 people are affected worldwide annually.[16] Cooking does not destroy the toxin.

Cats have been poisoned by consuming discarded toxic fish. Cats have sometimes been used in bioassays because the toxin consistently causes vomition in cats.[16]

C. POISONS

Ciguatoxin is a heat-stable, lipid-soluble compound with a molecular weight of approximately 1500. This toxin is distributed throughout the fish. Maitotoxin is also heat stable and, though it is water soluble, it remains in the intestine.[17]

TABLE 3
Fish Poisons

I. Poisoning caused by ingestion of preformed toxins in fish
   A. Ichthyosarcotoxic fish — the toxin is located in the musculature, viscera, and skin. The toxins are believed to be small molecular structures and are generally heat stable and not destroyed by gastric juices.
      1. Ciguatoxin group — tropical reef fish
         a. Toxins
            Ciguatoxin
            Maitotoxin
            Brevetoxin a and b
            Okadaic acid
            35S-Methylokadaic acid
         b. Representative ciguatoxic fish
            Scomberomorus commersoni — Spanish mackerel
            Sphyraena barracuda — barracuda
            Lutjanus bohar — red snapper
            Variola louti — grouper
            Epinephelus fuscoguttatus — grouper
            Seriola dumurili — amber jack
            Anguilla japonica — Japanese eel
      2. Tetrodotoxin group — Families — *Tetraodontidae* and *Diodontidae*
            Arothron nigropunctatus — black-spotted puffer
            Diodon histrix — porcupine fish
      3. Histamine (scombroid fish) poisoning — scombroid poisoning is generally caused by spoilage of scombroid fish (family — *Scombridae* — tunas).
            Euthynnus pelamis — skipjack tuna
            Thunnus alalunga — albacore tuna
   B. Ichthyootoxic fish — poisonous gonads or roe
            Scorpaenichthys marmoratus — cabezon
            Stichaeus grigorjewi — northern blenny
            Lepisosteus spp. — gar
   C. Ichthyohemotoxic fish — these species have poisonous blood serum.
            Anguilla anguilla — common European eel
            Conger conger — conger eel
            Muraena helena — moray eel

II. Ichthyochrinotoxic fish — these fish produce their own toxin in specialized glands, but they are not associated with a venom apparatus. The toxic glandular contents are secreted into the water for the purpose of warning or repelling would-be predators.
   A. Representatives
            Lampetra fluviatilis — river lamprey
            Muraena helena — moray eel
            Arothron meleagris — puffer fish
            Opsanus tau — oyster toad fish

III. Venomous (acanthotoxic) fish — produce a venom and have an apparatus to inject the venom.
   A. Sharks and rays
            Heterodontus francisci — horn shark
            Squalus acanthias — spiny dog fish

TABLE 3 (continued)
Fish Poisons

		Dasyatis pastinaca — stingray
		Aetobatus marinari — spotted eagle ray
		Urolophus halleri — round stingray
B.	Catfish	
		Arius lineatus — saltwater catfish
		Plotosus lineatus — Oriental catfish
		Bagre marinus — sea catfish
C.	Weever fish — family — *Trachinidae*	
		Trachinus spp. — greater and lesser weever fish
D.	Scorpion fish — family — *Scorpaenidae*	
		Pterois volitans — zebra fish, turkey fish
		Scorpaena guttata — scorpion fish
		Synanceja horrida — stonefish
E.	Toad fish — family — *Batrachoididae*	
		Opsanus tau — oyster toad fish
		Thalassophryne maculosa — toad fish
F.	Stargazers — family — *Uranoscopidae*	
		Uranoscopus spp. — stargazers
G.	Rabbit fish	
		Siganus vulpinus — rabbit fish
H.	Surgeon fish	
		Acanthurus triostegus — convict surgeon fish
		Prionurus microlepidotus — surgeon fish

D. CLINICAL SIGNS

1. Humans

Clinical descriptions of ciguatoxicosis have varied, possibly because of variation in the size of the dose ingested and because more than one toxin may be involved. Ciguatoxicosis is rarely fatal, but may produce significant morbidity. Paresthesia (burning or tingling sensation of the extremities and orofacial area) is a prominent symptom and is used as a differential for other types of fish poisoning or mild gastroenteritis. Other neurotoxic-related signs include myalgia, ataxia, pruritus, vertigo, dizziness, extreme fatigue, headaches, transient blurred vision or blindness, sweating with chilling, muscular numbness, and auditory hallucinations. Neurologic signs may persist for months or even years.[16] An interesting phenomenon is that, sometimes, later ingestion of alcoholic beverages along with nontoxic fish may induce a recurrence of pruritus. Correlating this sign and previous ingestion of ciguatoxin may be difficult.[16]

Gastroenteric-related signs include watery diarrhea, nausea, vomiting, and colic. These signs usually persist for 1 to 2 days, after which the victim is left in a weakened condition for another 2 to 7 days.[16] Cardiovascular-related signs are less common, but may include sinus bradycardia with hypotension and cyanosis, and occasional tachycardia with arrhythmias.

Even though rarely fatal, acute cases of ciguatoxicosis may result in death caused by paralysis of the muscles required for respiration. Also, in chronic forms of the condition, dehydration associated with vomiting and diarrhea may be severe enough to cause death, particularly in children not provided with medical care.[16]

2. Animals

Cats are particularly sensitive to ciguatoxin. Severe poisoning may be prevented by the vomition produced soon after ingestion of toxic fish.

E. DIAGNOSIS

No laboratory tests have been developed that can confirm a diagnosis of ciguatoxicosis. A history of fish consumption plus the clinical signs are the only criteria for diagnosis.

F. TREATMENT

As there is no antidote for ciguatoxin poisoning, symptomatic and supportive care should be instituted. Numerous herbal, homeopathic, and medical treatments have been advocated.[16,18]

G. PROGNOSIS

Usually, ciguatoxicosis is mild and short-term, especially when appropriate medical care is given.

H. PREVENTION

Ciguatoxicosis usually occurs in localized areas. Native inhabitants avoid eating fish inhabiting certain reefs. No particular fish can be identified as being toxic or nontoxic.

VI. TETRODOTOXIN POISONING

A. IDENTIFICATION

Representative species of fish containing tetrodotoxin are listed in Table 3. Marine puffer fish (toad fish, globe fish, toado, swell fish, porcupine fish (Figure 2), and balloon fish) are found throughout the world, usually in warm waters.[10] Names are derived from the behavior of these fish that balloon up to increase size when threatened. Fisher people who recognize these fish commonly kill them or throw them back into the water.

B. CONDITIONS OF POISONING

1. Human

In Japan, the flesh of certain species of puffer fish (fugu) is highly prized. Fugu must be prepared by specially trained cooks to minimize the hazard; even so, in the 20-year period from 1955 to 1975, 3000 people were poisoned by fugu in Japan; 51% of those individuals died.[17]

One reason for the popularity of fugu is a pleasant tingling sensation, apparently produced by traces of the toxin. Some dishes are served raw, in thin slices, which does not increase danger, as the toxin is not destroyed by

FIGURE 2. A porcupine fish, *Diodon histrix*, family Diodontidae. (Photo by the author, courtesy Steinhart Aquarium, San Francisco, CA.)

cooking.[17] Captain Cook, the English explorer, nearly died of tetrodotoxin poisoning in New Caledonia in 1774.[17]

2. Animal

Dogs and cats may be poisoned by scavenging discarded fish or viscera. In one instance, a family cat and a boy were both fatally poisoned; in another, a pet crow was killed.

C. POISON

Tetrodotoxin has been isolated and its complicated structural formula identified. The empirical formula is $C_{11}H_{17}O_8N_3$; the molecular weight 319.3. Tetrodotoxin is a potent vasopressor and neurotoxin, blocking the excitability of nerves by interfering with the sodium channel.

Tetrodotoxin is primarily concentrated in the liver, ovaries, intestine, and skin, but the flesh easily becomes contaminated while the fish is being cleaned.[17] Tetrodotoxin has also been isolated from such diverse animal groups as Central American frogs of the genus *Atelopus* (skin), California newt (salamander), *Taricha californicus* (skin glands), an octopus, *Hapalochlaena maculosa* (posterior salivary glands), and the Pacific goby, *Gobius criniger*.[17]

1. Toxicity

The lethal toxicity of crystalline tetrodotoxin, administered orally, has been determined in the following animals: mouse, 180 μg/kg; rat, 147 μg/kg; rabbit,

200 μg/kg; and dog, 70 μg/kg. Comparable toxicity for toxin administered intravenously is as follows: mouse, 8 μg/kg; rat, 10 μg/kg; rabbit, 2 μg/kg; cat, 2 μg/kg; and dog, 0.3 μg/kg.[17]

D. CLINICAL SIGNS

1. Humans

Signs of poisoning occur within 10 to 45 min after ingestion of the fish, beginning with numbness around the mouth that spreads to include the tongue, face, and other areas of the skin. Signs of motor paralysis develop early, shown by incoordination and slurred speech, which may give the appearance of drunkenness. The patient remains conscious until near death, but a generalized paralysis with loss of voice, dyspnea, and hypotension develops. Grave signs include severe hypotension, respiratory paralysis, and hypoxia. Death may follow as quickly as 17 min after ingestion of the fish.[17]

2. Animals

The syndrome seen in experimental animals is similar to that observed in people. Cats developed gross mydriasis, flaccid paralysis, tachycardia, and depressed respiration, but recovered in 24 h.[17] Cats may vomit early in the course of poisoning, reducing the ingested dose of tetrodotoxin.

In one clinical case, a pet cat became paralyzed in the rear limbs within 2 h after ingestion and was found dead the next morning.

Birds are also susceptible to the toxin. An Australian pet crow (magpie) began to stagger moments after ingestion of a puffer fish. In a few minutes, it fell with its wings twitching and died.[17]

E. DIAGNOSIS

The rapid onset of neurologic signs plus a history of puffer fish ingestion is sufficient to make a diagnosis. No laboratory test can confirm a suspected diagnosis. Lesions are minimal at necropsy, other than agonal hypoxic ecchymoses on serosal surfaces and pulmonary edema.

A differential diagnosis should include tick paralysis and ciguatoxin poisoning.

F. TREATMENT

As there is no antidote, symptomatic and supportive care are recommended. Controlled ventilation is crucial in severe poisoning. Intravenous fluids are required to relieve hypotension.

G. PROGNOSIS

Grave, unless immediate hospital management can be provided.

H. PREVENTION

Avoid ingestion of toxic fish. Consuming fugu, including at first-class restaurants, is risky.

FIGURE 3. A zebra fish (turkey fish, lion fish) *Pterois volitans,* Family Scorpaenidae. (Photo by the author, courtesy Steinhart Aquarium, San Francisco, CA.)

VII. VENOMOUS FISH

A. IDENTIFICATION AND BIOLOGY

Table 3 lists representative species of venomous fish. Numerous species have venomous spines associated with the dorsal or pectoral fins. A few species, such as the zebra fish *(Pterois volitans)* shown in Figure 3, and the stone fish *(Synanceja verrucosa)* have an elaborate array of venomous spines associated with pectoral, dorsal, pelvic, and anal fins. Venomous fish and rays inhabit most of the oceans of the world but are concentrated in tropical waters. A few venomous rays inhabit fresh water.[11]

B. CONDITIONS OF POISONING

Commercial fisher people recognize venomous species in their locale, but novices may be envenomated when trying to remove a fish from a net or hook. Persons wading in shallow water near sandy beaches may be impaled by a stingray. Scuba divers may invade the territory of some of the more aggressive weever fish and scorpion fish, which may actually attack them. The author was somewhat alarmed when surrounded by six zebra fish (turkey fish, *Pterois volitans)* (one of the aggressive species) while skin-diving on a coral reef off Fiji. Fortunately, they remained quiet and undisturbed, seemingly unaware of the humans in their vicinity.

Public and private aquaria are now commonplace. Ornate species, such as zebra fish, make an excellent display. As long as aquarists are knowledgeable about these fish, accidents may be prevented. The amateur fish hobbyist is another matter. The author once received a call from a local physician who was administering emergency care to a pet shop owner who had just been envenomated by an oriental catfish, *Plotosus lineatus*. He had used a dip net to transfer the fish from a stock tank to a container for a customer. The fish flopped out of the net and the man unthinkingly reached out and caught it to prevent its injury by hitting the floor. The immediate, sharp pain was so intense it caused him to fall to the floor. The amateur fish fancier who acquires such a fish is at risk.

C. VENOM

Fish venoms have not been studied as intensively as other venoms, but it is known that they contain proteins and nonprotein substances that may be highly toxic. Many of the venoms are heat labile, a factor useful in first aid and therapy.

D. CLINICAL SIGNS

The sting of a zebra fish (turkey fish) produces immediate intense pain, severe persistent edema, contusion-like discoloration, and vesication. The wound site may necrose and slough. Alarming systemic responses include dyspnea and shock. Fatalities have occurred. Persons who are stung while in deep water may be impaired sufficiently to be in danger of drowning if help is not nearby.

Stone fish envenomation causes a similar local reaction, with the addition of numbness and paralysis of the limb involved. Systemic signs include generalized weakness, profuse sweating, nausea, vomiting, diarrhea, headache, delirium, cardiac arrhythmias, dyspnea, and convulsions, with death occurring within 6 h.[14]

The stiff spines associated with the fins of many nonvenomous fish may cause lacerations when the fish is handled. Mucous secretions on the skin surface of such fish may enter the wound and cause local inflammation, similar to mild effects of envenomation.

E. TREATMENT

1. Human

Wounds should be examined to remove any remaining part of a spine. Application of a tourniquet in the first aid and treatment of venomous bites and stings is rarely recommended; however, this is a situation where such use appears to be justified. A tourniquet should be applied proximal to the wound and loosened every 5 min. The involved limb should be immersed in hot water (50 to 60°C, 122 to 140°F) for 30 to 90 min.[14] The rationale for this form of therapy is based on the thermolability of the venom. A thermometer should be used to determine water temperature in order to avoid scalding the patient.[14]

F. DIAGNOSIS

Response to fish envenomation is immediate. The victim may not know what species of fish or ray caused the envenomation, but the occurrence is obvious. Victims should be monitored for several hours to avoid the disastrous consequences of severe systemic reactions.

G. PROGNOSIS

Thousands of venomous fish stings occur each year around the world. Fatalities are rare. The pet store owner envenomated by the catfish was incapacitated by violent pain, but returned to normal a few hours following administration of meperidine.

H. PREVENTION

Venomous fish should not be allowed in the hobbyist trade. Danger of envenomation is also inherent with careless wading in shallow tropical waters or when reef walking.

VIII. MISCELLANEOUS VENOMOUS AND POISONOUS MARINE ANIMALS

Some of these less-important poisonous or venomous species are mentioned to provide an overview. Only rarely are domestic animals involved in poisoning. However, wild animals may be poisoned, though documentation of such occurrences is lacking.

A. SPONGES — *PORIFERA*

Sponges are widespread throughout the world, inhabiting the sea floor. There are over 4000 species of sponges, some of which have been important in commerce. Living sponges produce a toxic secretion that is unpalatable to most other animals. A characteristic of sponge anatomy is incorporation of sharp calcareous or siliceous spicules that may lacerate the skin, allowing the toxic secretion to produce a painful dermatitis.

The toxins found in sponges have been characterized and named, usually according to the species from which they were extracted.[19]

B. MARINE SNAILS — *MOLLUSCA*

Besides causing paralytic shellfish poisoning (from a preformed toxin acquired from dinoflagellates), some mollusk species also produce a venom used in food acquisition[20] (see Table 2). The most important of the venomous mollusks are the cones. In these species, the radula, the rasping organ used to scrape food from surfaces, has been modified to project tiny venomous darts at prey species.[21]

The w-conotoxins in cone venom are unique in that they inhibit neuronal calcium channels, causing immediate paralysis of the victim. Human envenomation may be lethal as well.[22]

C. OCTOPUS

Only one small species, the blue-ringed octopus, *Octopus maculosus,* has been recorded as inflicting lethal bites in humans.[9] This species inhabits ocean waters in the Indo-Pacific, Australian, Japanese, and Indian areas. A few other octopus species and their relatives, the cuttle fish, have been recorded as being slightly venomous.[6] The venom is similar in chemical structure and effect to tetrodotoxin.[22] Animal envenomation has not been reported.

D. ANNELID WORMS

Some annelid marine worms produce a paralytic biotoxin that is injected into invertebrate prey species. Fisher people use these worms as a favorite live bait for catching fish.[23] In addition to the venom apparatus associated with the mouth parts, they also have setae or bristles that may cause a localized dermatitis in humans.[24]

The effects of the biotoxin (nerestoxin) produced by *Lumbriconereis heteropoda* were observed over 40 years ago by fisher people who used the worms as live bait. The toxin is located in the skin of the worm. It was observed by bait shopkeepers that insects that fed upon the worms died or became incapacitated, and dead worms were hung in strategic places in Japanese bait shops to control insects. The toxin has been isolated, synthesized, and is now being marketed in Japan as an insecticide particularly effective against rice stem borer larvae.[23]

Nerestoxin acts primarily as a nicotinic cholinergic receptor antagonist at the vertebrate neuromuscular junction. Humans have developed dyspnea, headaches, and vomiting after handling the worms.[23]

Another annelid, *Glycera convoluta* (blood worm), is used by New England fisher people. This species has venom glands associated with fangs on the mouth parts. The toxin is of interest to the toxinologist, but is of no consequence in domestic animal poisoning.

E. ECHINODERMS

Sea urchins, starfish, sea cucumbers, and nudibranchs are prominent fauna of the sea floor. Starfish and sea cucumbers are primarily toxic only to their prey species. Sea urchins present more of a hazard to humans who may come into physical contact with the spines, some of which have poisonous sacs associated with the tip. Spine penetration in humans causes pain, edema, and erythema. In a few instances, multiple penetrations have resulted in numbness, paralysis, and bronchospasm.[23]

F. REPTILES

Sea snakes are discussed in the Australian reptile chapter (Chapter 17). The flesh of sea turtles has been reported to be poisonous in rare instances. As these turtles are herbivorous, it is likely that toxicity is the result of ingestion of a particular poisonous plant.[6]

G. MAMMALS

Cetaceans have provided food for native populations in many countries of the world; however, clinical reports indicate that consumption of the liver and flesh of some cetaceans has caused mortality in humans (Table 2).[6] The nature of the toxin is unknown in most cases. The syndrome varies, but includes headache, gastrointestinal upset, flushing of the face, facial edema, desquamation of the skin, and liver impairment.[6]

The liver and kidneys of seals, sea lions, walrus, and polar bears may contain high levels of vitamin A. Polar bears probably acquire vitamin A from preying on seals. Native people discard these organs, but uninformed Arctic travelers have developed symptoms of hypervitaminosis A by ingesting moderate to large amounts of the liver or kidneys of polar bears.[1,6]

The clinical syndrome of polar bear poisoning begins 2 to 5 h after ingestion of liver or kidney. Signs include headache, nausea, vomiting, diarrhea, abdominal pain, dizziness, drowsiness, irritability, weakness, muscle cramps, visual disturbances, and collapse.[6] Headaches become intense during the next 8 h and are exacerbated by lying down. Desquamation of the face, arms, legs, and feet commonly occurs within a day or two. Fatalities are rare and the alarming signs usually disappear within 24 h.[6]

The syndrome is similar to that of clinical hypervitaminosis A caused by excessive ingestion of therapeutic vitamins. Signs may develop following a single ingestion of a megadose (several million units in an adult) of vitamin A and include violent headaches, nausea, vomiting, drowsiness, polyarthralgia, and localized desquamation of the epithelium of the skin about the lips. Neurologic signs are apparently associated with an abrupt and marked increase in cerebrospinal fluid pressure. The signs generally abate within 48 h.[6]

REFERENCES

1. **Halstead, B. W. and Vinci, J.,** Biology of poisonous and venomous marine animals, in *Handbook of Natural Toxins, Vol. 3. Marine Toxins and Venoms,* Tu, A. T., Ed., Marcel Dekker, New York, 1988, 1.
2. **Carmichael, W. W.,** Toxins of freshwater algae, in *Handbook of Natural Toxins, Vol. 3. Marine Toxins and Venoms,* Tu, A. T., Ed., Marcel Dekker, New York, 1988, 121.
3. **Baden, D. G.,** Public health problems of red tides, in *Handbook of Natural Toxins, Vol. 3. Marine Toxins and Venoms,* Tu, A. T., Ed., Marcel Dekker, New York, 1988, 259.
4. **Shimizu, Y.,** The chemistry of paralytic shellfish toxins, in *Handbook of Natural Toxins, Vol. 3. Marine Toxins and Venoms,* Tu, A.T., Ed., Marcel Dekker, New York, 1988, 63.
5. **Evans, M. H.,** Block of sensory nerve conduction in the cat by mussel poison and tetrodotoxin, in *Animal Toxins,* Russell, F. E. and Saunders, P. R., Eds., Pergamon Press, Oxford, England, 1967, 97.
6. **Halstead, B. W.,** *Poisonous and Venomous Marine Animals of the World,* Revised ed., Darwin Press, Princeton, NJ, 1978.
7. **Walker, M. J. A.,** Coelenterate and echinoderm toxins: mechanisms and actions, in *Handbook of Natural Toxins, Vol. 3. Marine Toxins and Venoms,* Tu, A. T., Ed., Marcel Dekker, New York, 1988, 279.
8. **Lane, C. E.,** Recent observations on the pharmacology of *Physalia* toxin, in *Animal Toxins,* Russell, F. E. and Saunders, P. R., Eds., Pergamon Press, Oxford, England, 1967, 131.

9. **Caras, R.,** Stingers in the sea: the coelenterates, in *Venomous Animals of the World,* Prentice-Hall, Englewood Cliffs, NJ, 1974, 13.
10. **Burns, T. A.,** Ichthyootoxic fish, in *Handbook of Natural Toxins, Vol. 3. Marine Toxins and Venoms,* Tu, A. T., Ed., Marcel Dekker, New York, 1988, 227.
11. **Castex, M. W.,** Fresh water venomous rays, in *Animal Toxins,* Russell, F. E. and Saunders, P. R., Eds., Pergamon Press, Oxford, England, 1967, 167.
12. **Halstead, B. W. and Courville, D. A.,** *Poisonous and Venomous Marine Animals of the World,* Vol. I, II, and III, United States Government Printing Office, Washington, D.C., Vol. I, 1965; Vol. II, 1967; Vol. III, 1970.
13. **Larson, E., Lalone, R. C., de Villez, E. J., and Siman, R., Jr.,** Physiological and pharmacological studies on extracts of the western Atlantic puffer *Spheroides maculatus,* in *Animal Toxins,* Russell, F. E. and Saunders, P. R., Eds., Pergamon Press, Oxford, England, 1967, 187.
14. **Maretic, Z.,** Fish venoms, in *Handbook of Natural Toxins, Vol. 3. Marine Toxins and Venoms,* Tu, A. T., Ed., Marcel Dekker, New York, 1988, 445.
15. **McFarren, E. F.,** Differentiation of the poisons of fish, shellfish and plankton, in *Animal Toxins,* Russell, F. E. and Saunders, P. R., Eds., Pergamon Press, Oxford, England, 1967, 85.
16. **Withers, N. W.,** Ciguatera fish toxins and poisoning, in *Handbook of Natural Toxins, Vol. 3. Marine Toxins and Venoms,* Tu, A. T., Ed., Marcel Dekker, New York, 1988, 31.
17. **Sutherland, S. K.,** *Australian Animal Toxins,* Oxford University Press, Melbourne, Australia, 1983.
18. **Auerbach, P. S.,** Clinical therapy of marine envenomation and poisoning, in *Handbook of Natural Toxins, Vol. 3. Marine Toxins and Venoms,* Tu, A. T., Ed., Marcel Dekker, New York, 1988, 493.
19. **Stonik, V. A. and Elyakov, G. B.,** Structure and biologic activities of sponge and sea cucumber toxins, in *Handbook of Natural Toxins, Vol. 3. Marine Toxins and Venoms,* Tu, A. T., Ed., Marcel Dekker, New York, 1988, 107.
20. **McMichael, D. F.,** Mollusks, in *Venomous Animals and Their Venoms,* Vol. 3, Bucherl, W. and Buckley, E. E., Eds., Academic Press, Orlando, FL, 1971, 373.
21. **Endean, R., Izatt, J., and McColm, D.,** The venom of the piscivorous gastropod *Conus striatus,* in *Animal Toxins,* Russell, F. E. and Saunders, P. R., Eds., Pergamon Press, Oxford, England, 1967, 137.
22. **Olivera, B. M., Gray, W. R., and Cruz, L. J.,** Marine snail venoms, in *Handbook of Natural Toxins, Vol. 3. Marine Toxins and Venoms,* Tu, A. T., Ed., Marcel Dekker, New York, 1988, 327.
23. **Kem, W. R.,** Worm Toxins, in *Handbook of Natural Toxins, Vol. 3. Marine Toxins and Venoms,* Tu, A. T., Ed., Marcel Dekker, New York, 1988, 353.
24. **Halstead, B. W.,** Venomous echinoderms and annelids, in *Venomous Animals and Their Venoms,* Vol. 3, Bucherl, W. and Buckley, E. E., Eds., Academic Press, Orlando, FL, 1971, 419.

PART III: Arthropoda

Chapter 4

ANTS

I. ETIOLOGY

There are 10,000 or more species of ants.[1-3] Some only bite, others only sting, and a third group both bite and sting.[4,5] Envenomation and death of pets, livestock, and wild animals occur from multiple stings from the third group in the New World (Tables 1 and 2). Species of fire ants native to the U.S. are local pests and may cause illness in debilitated or neonate animals. Harvester ants are pugnacious, vicious stingers and may cause mild envenomation.

The red imported fire ant is the most important of the ants as far as animal envenomation is concerned. Worker ants (wingless, sterile females) vary from 1.6–5.8 mm in length. Body color is red to reddish brown. Nonworking males and females are slightly larger. As many as 30 colonies per acre have been found. Each colony is a hard encrusted mound, 25 to 65 cm in diameter, and may contain 30,000 to 100,000 workers. Fire ants colonize pastures, corrals, and farm yards.[4,5]

Red imported fire ants eat roots, stems, buds, and fruits of plants. Seedling trees may be girdled. These ants are also predators of insects and debilitated or immobile mammals, birds, and reptiles.[6]

A. DISTRIBUTION

Fire ants are distributed throughout the neotropics. The red imported fire ant is indigenous to central Brazil, but entered the U.S. in the 1940s and is now a serious pest throughout the southeastern states and west as far as Texas (Figure 1).[4-6] Control methods have been unsuccessful in halting the spread of this pest, and suitable habitat for the species exists across the southern tier of states, west to California, and north into Nevada, Utah, Oklahoma, Arkansas, and Tennessee, and as far north as New Jersey and New York on the east coast.[6]

Harvester ants and native fire ants are found throughout the U.S. Although ants are found throughout the world, envenomation of animals is a problem only in North, Central, and South America.

B. BIOLOGY

All ants live in communities composed of a few to a million ants.[7] The social structure of ant colonies has piqued the interest of biologists for thousands of years. The division of labor is precise in many ant colonies, particularly those species involved in animal envenomation. The fundamental drive of ants is to protect and feed the colony (population). Numerous strategies are employed.

The majority of ants in a nest are workers (always infertile, wingless females). The stinger apparatus evolved from the ovipositor and as such is present only in females. Workers do all the work in the colony except the laying

TABLE 1
Venomous Ants of the U.S.

Common name	Scientific name	Geographical location
California harvester ant	*Pogonomyrmex californicus*	Texas, Utah California
Western harvester ant	*P. occidentalis*	Kansas, Colorado, Wyoming, Arizona, Nevada
Red harvester ant	*P. barbatus*	Kansas, Louisiana to Utah and California
Red imported fire ant	*Solenopsis invicta*	Figure 1
Black imported fire ant	*S. richteri*	Mississippi
Common fire ant	*S. geminata*	Texas to South Carolina and Florida
Southern fire ant	*S. xyloni*	California to South Carolina and Florida

TABLE 2
Classification of Ants Mentioned in Chapter 4

Class — *Insecta*
- Order — *Hymenoptera*
  - Infraorder — *Aculeata*
    - Family — *Formicidae*
      - Subfamily — *Formicinae*
        - *Formica* spp. — field ants
      - Subfamily — *Ponerinae*
        - *Paraponera clavata*
      - Subfamily — *Pseudomyrmecinae*
        - *Pseudomyrmex mexicanus*
      - Subfamily — *Myrmicinae*
        - *Solenopsis invicta* — red imported fire ant
        - *Solenopsis* spp. — fire ants
        - *Myrmica* spp.
        - *Pogonomyrmex* spp. — harvester ants

of eggs. Workers may be subdivided into medium-sized ants and a few giant ants (called soldiers). The queen is a sexually mature female that is inseminated during her winged nuptial flight by males that die soon thereafter.[7] The queen then burrows into soil or other debris, loses her wings, and either pauses for a time or immediately begins to lay eggs.[7] All workers in a nest are descended from a single queen.[7]

Ants have powerful mandibles used for obtaining food.[8] Even nonvenomous ants may inflict painful bites. The venom glands in ants may be the source of alarm pheromones, trail pheromones, sex pheromones, attractant-recruitment pheromones, aggregation pheromones, and recognition pheromones, all important in communication.[1,9]

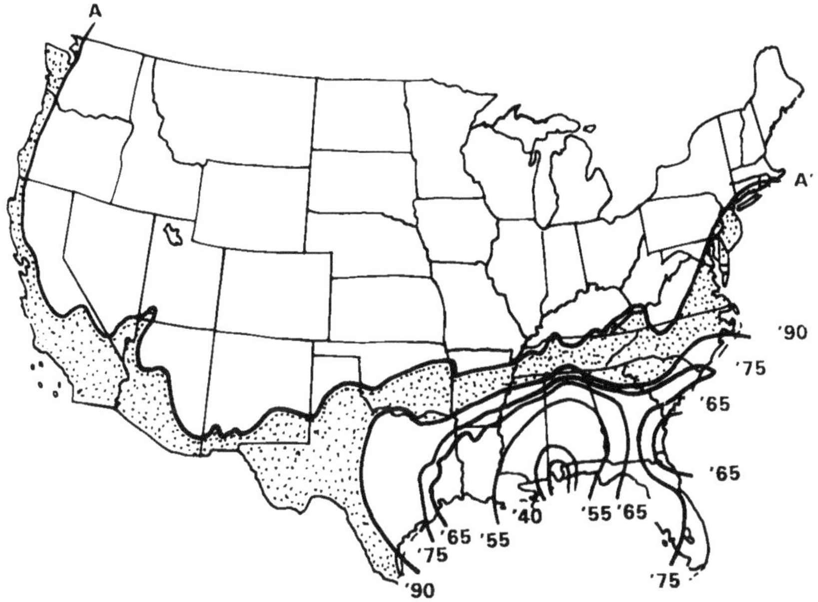

FIGURE 1. Map illustrating the spread of the red imported fire ant in the U.S. The stippled area beneath the line A-A′ indicates potential suitable habitat. (Adapted from Harwood, R. F. and James, M. T., *Entomology in Human and Animal Health,* 7th ed., Harwood, R. F. and James, M. T., Eds., Macmillan, New York, 1979, 430.)

II. CONDITIONS OF POISONING

Healthy adult animals are minimally affected if they can move away from a colony of ants. Neonates, juveniles, and animals that are disabled and recumbent may be attacked and killed if in the vicinity of multiple ant colonies. Nesting bird chicks have been victimized. The venoms of all the fire ants are toxic, but the imported red fire ant's aggressive behavior makes envenomation more likely.

III. VENOM

Many species of ants produce venom, but only a few species have behaviors or sufficiently toxic venom to place vertebrate animals at risk. Ants employ venom in *offense* to obtain food (predation); *defense* (primary function of most ant venoms); and *chemical communication* (alarm pheromones are important in social species to call fellow ants to attack an enemy).[1,10]

Formic acid was identified in ant venom over 300 years ago.[1,11] Unfortunately, it was subsequently falsely assumed that all the effects of ant stings (and of all Hymenoptera) were caused by formic acid. Modern toxinologists have found ant venoms to be complex substances and species specific, just as are the venoms of snakes.

Ant venoms vary considerably from group to group.[12] The venom in the formicine group (*Formica* sp.) is primarily composed of formic acid. These ants bite with powerful mandibles, arch the abdomen under the body, and spray the venom onto the wounds. When venom contacts the dermis, there is immediate pain. In addition, the venom is also corrosive, causing cellular necrosis.[4] Nevertheless, these ants are of little consequence in serious animal envenomation.

A second venom group is composed primarily of proteinaceous substances (enzymes and peptides). Most of the venomous ants (*Ponerinae, Pseudomyrecinae,* and *Myrmicinae*) produce this type of venom. As in bee venom, phospholipase and hyaluronidase predominate, but numerous other enzymes are present.[1] A protein in the venom of *Pogonomyrmex barbatus* is a potent hemolysin.[1,13] Other biologically active compounds isolated from ant venoms include histamine, 5-hydroxytryptamine, acetylcholine, norepinephrine, and dopamine.[1]

When venom in the proteinaceous group is used to obtain food, the primary objective is paralysis of an invertebrate victim. In vertebrates, the neurotoxic action may cause convulsions, apnea, and rapid death.[14]

The venom of fire ants (*Solenopsis* spp.) is unique in that the majority of the venom consists of alkaloids (methyl-*n*-alkylpiperidines).[15] Alkaloids usually have been thought to be produced only by plants. Fire ant venom may also contain other nonproteinaceous, biologically active compounds.

IV. CLINICAL SIGNS

Pain is associated with the initial sting. A dog may yelp or a horse jump away from the area stung. Human victims quantitate the intensity and duration of pain.[1] The venom of the neotropical ant *Paraponera clavata* produces immediate, intense, even debilitating pain that may persist for 24 h. Sometimes numbness, vomiting, inflammation, and uncontrollable trembling accompany the pain.[1,16]

Harvester ants, *Pogonomyrmex* spp., also produce painful stings, but pain may not develop for 30 s and usually persists for 1 to 4 h. Harvester ants are considered the most aggressive and pugnacious of the native ants in the U.S.[4] Pain from a *Solenopsis* sting is not so intense and lasts only 2 to 5 min, but there is a prolonged local reaction.[1]

Fire ants first bite and, while grasping the victim, arch the back to thrust the stinger into the skin, sometimes in multiple locations while still attached at one bite site. The sting elicits immediate, sharp pain. Within a few minutes, an urticarial weal forms, which transforms into a vesicle and then into a pustule within 24 h. The pustule may persist for 3 to 8 d before rupturing and leaving a crust.[4]

Fire ants will devour newly hatched quail and poultry or enter pipping eggs.[5] Neonates of all species are at risk. Bites and stings of fire ants may cause focal necrotic ulcers of the cornea and conjunctiva of newborn calves.[17]

Multiple ant stings may result in systemic manifestations, including convulsions, apnea, shock, and rapid death. Anaphylactoid responses have not been reported in animals.

V. DIAGNOSIS

The clinical appearance of the sting is typical, but may be confused with other insect stings. A history of ant colonies in the vicinity of the animal is necessary for diagnosis. Ants may be observed crawling on the animal by an owner or caretaker.

VI. TREATMENT

Single ant bites do not require treatment. Multiple envenomation may result in a severe systemic reaction requiring intensive care for survival of the victim. Corticosteroids are indicated to diminish inflammatory and allergic responses. Initially, prednisolone sodium succinate at 5 to 10 mg/kg should be administered intravenously. Dexamethasone may also be administered at 0.2 mg/kg for an antiinflammatory effect or 2 to 4 mg/kg to counteract shock.

If anaphylaxis is diagnosed, epinephrine should be administered intravenously. The dosage for large animals is 0.01 mg/kg [3 to 5 ml of a 1:1000 (1 mg/ml) solution per 450-kg animal]. The dosage for small animals is 0.01 to 0.02 mg/kg of a 1:10,000 dilution (0.1 mg/ml). A 10-kg dog would require 1 ml solution.[18,19]

VII. PREVENTION

To date, no control programs have been able to eradicate or prevent the spread of the red imported fire ant. Other ant species may be controlled with insecticides. Control measures should not be attempted without consultation with specialists in ant control. It is particularly important to eradicate ant colonies in areas where neonates are housed.

REFERENCES

1. **Schmidt, J. O.,** Chemistry, pharmacology, and chemical ecology of ant venoms, in *Venoms of the Hymenoptera,* Piek, T., Ed., Academic Press, Orlando, FL, 1986, 425.
2. **Caras, R.,** Insects, in *Venomous Animals of the World,* Caras, R., Ed., Prentice-Hall, Englewood Cliffs, NJ, 1974, 88.
3. **Rathmayer, W.,** Hymenoptera: the aculeate wasps, ants and bees, in *Grzimek's Animal Life Encyclopedia,* Vol. 2, Grzimek, B., Ed., Van Nostrand Reinhold, New York, 1975, 421.

4. **Eberling, W.,** Ants, in *Urban Entomology*, Eberling, W., Ed., University of California, Division of Agricultural Sciences, Los Angeles, CA, 1975, 346.
5. **Harwood, R. F. and James, M. T.,** Ants, in *Entomology in Human and Animal Health*, 7th ed., Harwood, R. F. and James, M. T., Eds., Macmillan, New York, 1979, 430.
6. **Fowler, M. E.,** Zootoxins, in *Current Veterinary Therapy, 3. Food Animal Practice*, Howard, J. L., Ed., W. B. Saunders, Philadelphia, 1992. (in press).
7. **Markl, H.,** Superfamily *Formicoidea* (ants), in *Grzimek's Animal Life Encyclopedia*, Vol. 2, Grzimek, B., Ed., Van Nostrand Reinhold, New York, 1975, 441.
8. **van Marle, J.,** Morphology of the venom apparatus, in *Venoms of the Hymenoptera*, Piek, T., Ed., Academic Press, Orlando, FL, 1986, 17.
9. **Maschwitz, U. W. J. and Kloft, W.,** Morphology and function of the venom apparatus of insects — bees, wasps, ants and caterpillars, in *Venomous Animals and their Venoms*, Vol. 3, Bucherl, W. and Buckley, E. E., Eds., Academic Press, Orlando, FL, 1971, 1.
10. **Schmidt, J. O.,** Allergy to Hymenoptera venoms, in *Venoms of the Hymenoptera*, Piek, T., Ed., Academic Press, Orlando, FL, 1986, 509.
11. **Piek, T.,** Historical introduction, in *Venoms of the Hymenoptera*, Piek, T., Ed., Academic Press, Orlando, FL, 1986, 1.
12. **Blum, M. S.,** Poisonous ants and their venoms, in *Handbook of Natural Toxins, Vol. 2. Insect Poisons, Allergens and Other Invertebrate Toxins*, Tu, A. T. Ed., Marcel Dekker, New York, 1984, 225.
13. **Adrouny, G. A., Derbes, V. J., and Jung, R. C.,** Isolation of a hemolytic component of fire ant venom, *Science*, 130, 449, 1959.
14. **Rubens San Martin, P.,** The venomous ants of the genus *Solenopsis*, in *Venomous Animals and Their Venoms*, Vol. 3, Bucherl, W. and Buckley, E. E., Eds., Academic Press, Orlando, FL, 1971, 95.
15. **MacConnell, J. G., Blum, M. S., and Fales, H. M.,** Alkaloid from fire ant venom: identification and synthesis, *Science*, 168, 840, 1970.
16. **Oehme, F. W., Brown, J. F., and Fowler, M. E.,** Bees, wasps, hornets and ants, in *Casarett and Doull's Toxicology*, 2nd ed., Doull, J., Klassen, C. D., and Amdur, M. O., Eds., Macmillan, New York, 1980, 565.
17. **Joyce, J. R.,** Multifocal ulcerative keratoconjunctivitis as a result of stings by imported fire ants, *Vet. Med.*, 78, 1109, 1983.
18. **Meerdink, G. L.,** Bites and stings of venomous animals, in *Current Veterinary Therapy, VII. Small Animal Practice*, Kirk, R. W., Ed., W. B. Saunders, Philadelphia, 1980, 174.
19. **Meerdink, G. L.,** Bites and stings of venomous animals, in *Current Veterinary Therapy, VIII. Small Animal Practice*, Kirk, R. W., Ed., W. B. Saunders, Philadelphia, 1983, 155.

Chapter 5

BEES, WASPS, AND HORNETS

I. ETIOLOGY

Most of the aculeate *Hymenoptera* group of insects have a stinger apparatus used to paralyze prey species or for defense. Most are small or lack the power to thrust the stinger through a vertebrate skin and thus are of no consequence in envenomating either human beings or animals. Other species of bees, wasps, yellow jackets, and hornets inflict, at the least, a transient, painful sting and sometimes kill animals and people (Table 1). Bee and wasp stings caused approximately 23 human deaths per year in the U.S. during the 1950s.[1] During that same period, approximately 13 individuals per year died of snakebite. No statistics are available for fatal bee or wasp stings in animals.

Over 20,000 species of bees are distributed throughout the world.[2-4] The honey bee, *Apis mellifera,* is one of two domesticated insect species (the other is the silk worm, *Bombyx mori).* Native honey bees in tropical Southeast Asia were the source of the domestic honey bee.[3] When early American colonists missed honey as a sweetener, they imported European strains of the honey bee to Jamestown, Virginia, and the bees became known as European honey bees.

Each year in the U.S. alone, honey bees produce approximately 115 million kilograms (250 million pounds) of honey, valued at U.S. $200 million.[5] In addition, bees also manufacture 1.8 million kilograms of beeswax and several lesser products such as bee pollen, royal jelly, and bee venom.[5]

Although the production of honey is important, the pollination of flowers of agricultural crops is of even more value. More than 90 cultivated crops, having a combined value of $20 billion, are pollinated by bees. Strangely, this function of bees was not recognized or exploited until the late 1940s. Now, over 4 million hives or colonies are maintained, some in every state in the U.S. Approximately 1600 beekeepers migrate with their bees to pollinate flowers in the spring of the year.[5]

The African honey bee, *Apis mellifera adamsonii,* a native of tropical southern Africa, was introduced into Brazil in 1956 in the hope of improving the local European strain's ability to cope with warm weather. The bees readily adapted, but swarms escaped and have been moving northward at a rate of 150 to 200 miles per year (Figure 1). The African honey bee hybridizes with the domestic honey bee, from which it can be distinguished only by apicultural entomologists.[6]

A. STINGER APPARATUS

There are numerous variations in the stinger of aculeate *Hymenoptera.*[6-9] The stinger is a modification of the ovipositor apparatus that is common in insects.[4] Only female bees and wasps possess stingers.[4] Figure 2 is a schematic diagram of the honey bee stinger apparatus.[7] The primary venom is secreted

TABLE 1
Classification of the Bees, Wasps, and Hornets

Phylum — *Arthropoda*
Class — *Insecta*
Order — *Hymenoptera*
Suborder — *Aculeata* (Ln. aculeatus = equipped with sting)
Superfamily — *Vespoidea*
Family — *Vespidae*
Vespula (Paravespula) maculata, black hornet
Vespula maculifrons, yellow jacket
Dolichovespula maculata, white-faced hornet
Polistes spp., paper wasps
Superfamily — *Apoidea* (20,000 species)
Family — *Apidae* — bees
Apis mellifera, domestic honey bee
Apis mellifera adamsonii — African honey bee (killer bee)
Apis dorsata — giant Indian bee
Bombus californicus, bumble bee

from specialized cells in the acid glands and transported by tiny tubes to the venom sac reservoir. The bulb contains one-way valves that control the flow of venom during envenomation. An alkaline gland produces a secretion necessary to maximize the toxicity of the venom, which is added to the venom at the time of injection through the stylet and lancet. In the honey bee, the lancet is covered with retrograde barbs that attach the lancet to the skin. When honey bees sting a thick-skinned vertebrate, the retrograde barbs remain impaled in the victim and the entire stinger apparatus is pulled from the bee, resulting in its death.

B. BIOLOGY

Bees and wasps have evolved marvelous methods for housing and guarding their broods.[2-4] The most well-known example is the domestic honey bee with its hexagonal cells for encasing the egg within the hive. There are, however, solitary nesting species as well. Some wasp species are called "paper wasps" because they build their hive out of wood, chewed from wooden structures or branches, that is mixed with saliva and smeared into thin sheets of paperlike material, layer after layer. These paper hives may be constructed within the walls of buildings or attached to the eaves of buildings or tree branches.

The study of the biology of the social structure of bees and ants has occupied numerous entomologists throughout their careers, and continues to do so. Social species of bees and wasps are more likely to be involved in human and animal envenomation than are solitary species.

C. DISTRIBUTION

Envenomation by *Apis mellifera* may occur anywhere in the world except the high Arctic and Antarctic. Various species of wasps, hornets, yellow jackets, bumble bees, and other bees may be locally abundant and cause problems.

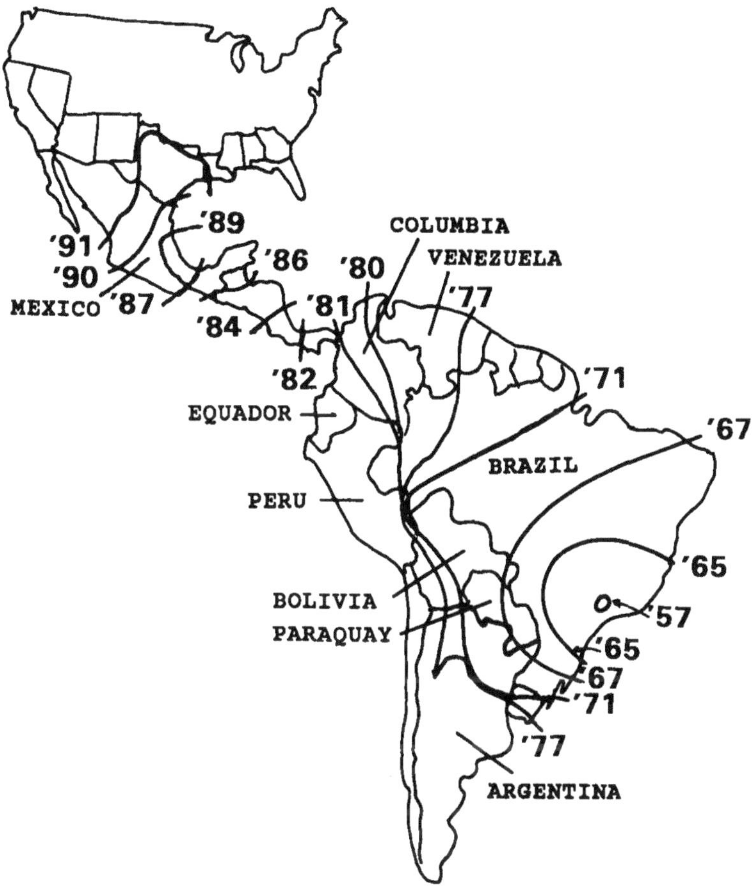

FIGURE 1. Diagram of the spread of the Africanized honey bee from Brazil to the U.S.

II. CONDITIONS OF ENVENOMATION

All domestic and wild animals may be envenomated by bees or wasps. Even animals as large as water buffalo, *Bubalus bubalis,* and elephant, *Elephas maximus*, have been killed by swarms of the giant Indian bee, *Apis dorsata.*[10,11] The effects of venom vary tremendously, depending on the species, the season of the year, the nutritional status, and age of the bee.[7]

A sting by a single bee or wasp rarely causes more than a transient, painful prick in an animal, in contrast to human medicine, where sensitive individuals may die peracutely from a single sting. However, a single sting has resulted in the death of a dog.[12] Any animal may collide with a single bee that will sting in defense.

Dogs have a propensity to snap at insects that hover over them. The swelling associated with a sting on the tongue or in the oral cavity may be life-

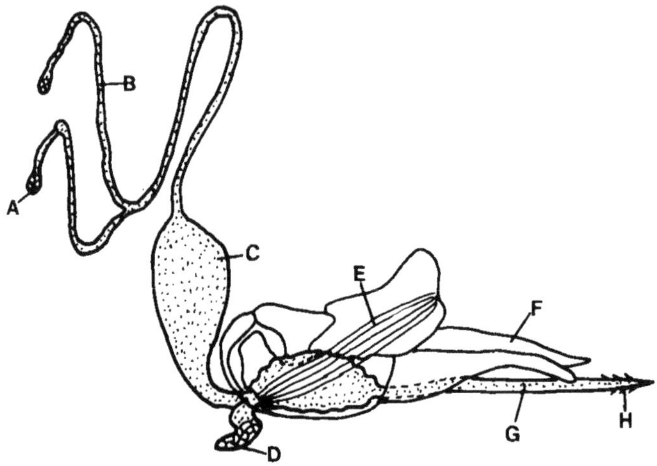

FIGURE 2. Diagram of the stinging apparatus of a honey bee, *Apis mellifera*. A, acid gland; B, venom duct; C, venom sac; D, alkaline gland; E, protractor muscle; F, sheath lobes; G, stylet; H, lancet (stinger). (Adapted from Croft, L. R., *Allergy to Bee Stings and Its Prevention*, Elmwood Books, Chorley, Lancashire, England, 1988.)

threatening. Yellow jackets are attracted to food and may become annoying at a picnic. Dogs that are fed outside may have to compete with yellow jackets for food and may be stung.

The author and a 5-year-old child were driving along a highway with the windows open, when the child cried out as if struck by an object. A painful weal developed on her arm within seconds. Later, a yellow jacket was found dead on the floor of the car. A pet may be similarly affected by an insect sucked into a vehicle through an open window.

Honey bees can inflict only a single sting, but an animal attacked by a swarm or hive of bees may sustain multiple stings and the cumulative envenomation may be lethal. Wasps and hornets either lack a barbed lancet or the barbs are small and do not prevent withdrawal of the lancet. Therefore, a single wasp may inflict multiple stings. Also, as dangerous species of aculeate *Hymenoptera* are usually social species, multiple stings are commonplace.

The Africanized honey bee presents a special case. Although its venom is no more toxic than that of the domestic honey bee, the African honey bee's aggressive behavior makes stinging more likely.[13] If a black flag were waved in front of a disturbed hive of bees, within a 30-s period there would be 10 stings by domestic honey bees and 85 by African bees. It requires about 20 s for the domestic honey bee to respond to a threat, and only 3 s for the African bee. Honey bees become calm in 2 to 3 min, but African bees remain agitated for 30 to 60 min. Africanized honey bees may pursue an enemy for 170 m, while the European honey bee loses interest in 27 m.[13] It should be noted, however, that not every researcher agrees that the African bee is so much more aggressive than the domestic bee.[6,14]

Livestock rarely succumb to the effects of bee stings except under unusual circumstances. If a horse is tied to a tree or shrub containing a nest of paper wasps and the nest is disturbed, the wasps may pour out of the nest and attack the horse. The cumulative effect of hundreds of stings may be lethal. Chacoan peccaries, *Catagonus wagneri,* were lethally stung by carrion-eating wasps, *Agelaia* sp., that built a nest in an earthen tunnel in which the peccaries slept.[15]

III. VENOM

Although the painful effects of bee envenomation have been known since antiquity, it has only been since the mid-1960s that scientific biochemical studies have been conducted on bee venom. For over 100 years, formic acid had been known to be the toxic chemical in ant venoms and it was erroneously assumed that all insect venoms were identical.[7,16] It is now known that even ant venom is complex, including other toxins as well as formic acid.

Bee venom contains two major toxic groups: allergens and non-allergens. Toxinologists have identified at least 18 allergens in whole bee venom.[17-19] The five primary allergens are phospholipase A_2, acid phosphatase, hyaluronidase, melittin, and an unidentified allergen (Ag-1).[7] A total of 55 different enzymes have been isolated from bee venom, but phospholipase A_2 and hyaluronidase are the two most important. Other components include histamine (may contribute to pain and the inflammatory response), dopamine, noradrenaline, amino acids, and volatile substances.[3] A number of reviews on the pharmacology and toxicology of bee and wasp venoms have been published.[20-25]

Bee venom allergy is more a human than an animal problem, but immediate hypersensitivity reactions do occur in animals and the phenomenon should be understood by those dealing with cases of animal insect envenomation.[16,26]

IV. CLINICAL SIGNS

A plethora of treatments has been recommended for insect stings. Myth and tradition are now being replaced by sound judgment based on the pathophysiologic effects of specific venins within a venom. Though animals may differ from humans in response to bee venom, the veterinarian is advised to understand the syndromes produced in humans.

1. Immediate hypersensitivity — anaphylactic reaction. This occurs only in humans or animals that have been previously sensitized to the allergens in the venom.
2. Local inflammatory response at the sting site. Initial pain (more severe in some individuals), erythema, and a weal or urticarial reaction, with a blanching of the site.
3. Systemic toxicity from the non-allergens in the venom. This effect may be delayed for minutes to hours.

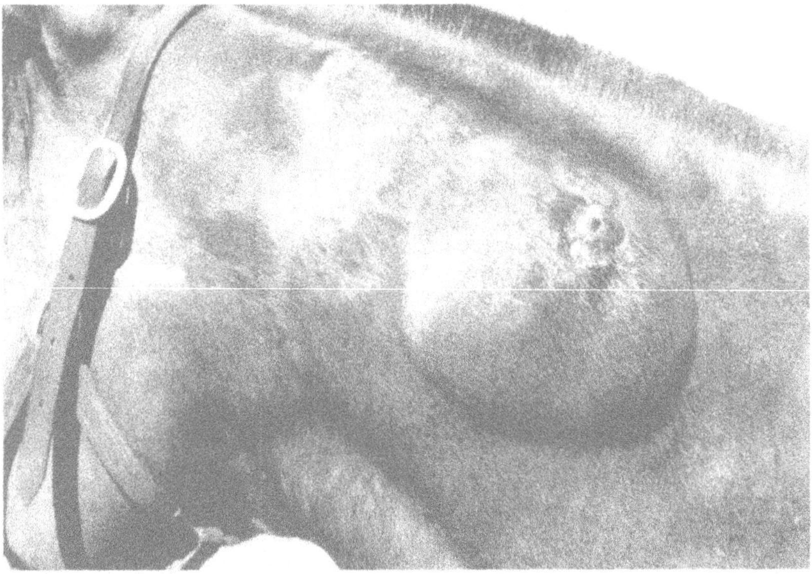

FIGURE 3. Insect bite reaction in a horse.

Hymenoptera hypersensitivity is an immune response in which the venom allergens react with cell-bound immunoglobulin E (IgE) to stimulate a massive release of histamine, leukotrienes, prostaglandins, and other chemotaxic factors.[16,17,26,27] The effects of these substances are contraction of smooth muscle plus dilation and increased permeability of small blood vessels, leading to bronchospasm, vomiting, urticaria, vascular collapse, and death.[7]

A. LIVESTOCK[28]

A local reaction consists of a swollen, edematous, and erythematous plaque (Figure 3).[29,30] Rarely, a honey bee stinger will be located. A tiny abscess may form at the injection site. Multiple stings may produce numerous urticarial weals (Figures 4 and 5) and cause severe systemic response as a direct toxic effect of quantities of venom. The classic anaphylactoid response of humans has not been documented in livestock.

Clinical signs of yellow-jacket envenomation of a group of swine included raised, circumscribed, vesicular-to-crusty lesions on the ears, snout, eyelids, anus, vulva, and scrotum.[31] A number of pigs became nervous, developed a cough, and exhibited increased thirst. Two males died, one during a convulsion. At necropsy, both had hemorrhagic gastric ulcers. A nest of yellow jackets was found in the pen. The remaining swine were moved, and recovered quickly.

Numerous livestock were killed by swarms of Africanized honey bees in Brazil, when the bees first began to spread, following their escape in the late 1950s. Reports of the "killer" potential of the bees were sensationalized in the

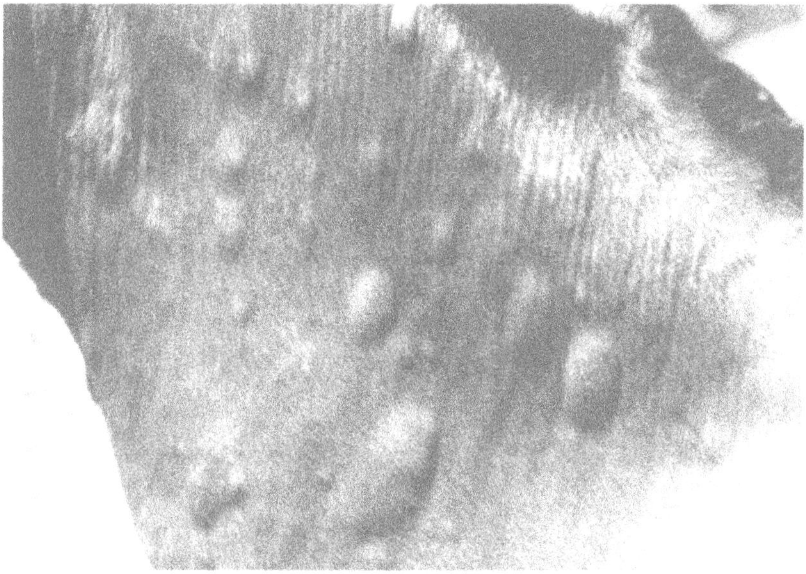

FIGURE 4. An urticarial reaction to insect envenomation in a horse.

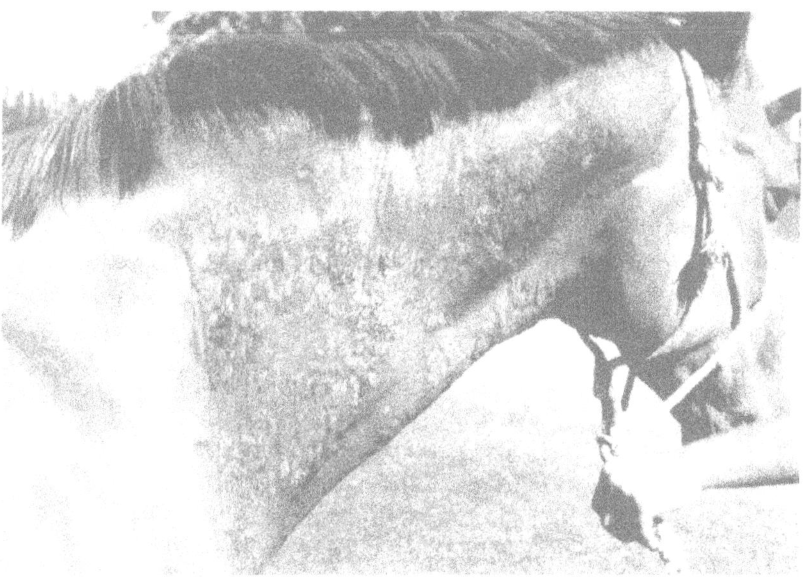

FIGURE 5. Secondary desquamation following an urticarial reaction in a horse.

press, continuing to the present time, but there is reason to believe that large-scale deaths have not been as common in recent years.[14]

The pain associated with multiple stings may cause excitement or frenzy in a horse, followed by tachycardia, diarrhea, hemoglobinuria, icterus, and prostration. If the stings are on the head near the nostrils or mouth, swelling may occlude air passages, causing dyspnea.

B. DOGS AND CATS[32-35]

Dogs may yelp at the initial sting. Cats are less likely to vocalize. Most victims of single stings are presented to the veterinarian with facial, aural, or periorbital edema.[36,37] If a dog snaps at a bee or wasp and the sting occurs in the oral or pharyngeal cavity, swelling may occlude the passageway, resulting in severe dyspnea.

Multiple bee stings may result in cardiovascular and respiratory collapse, with death ensuing in minutes. Pets may respond to intensive care therapy, but still develop severe signs of toxicity, as reported by Wysoke et al.[36] in South Africa. In the three cases they described, there were signs of anorexia and CNS (central nervous system) depression, with specific neural deficits of various nerves (facial and trigeminal cranial nerve paralysis, caudal ataxia). The three dogs developed anemia, with one exhibiting severe hemoglobinuria. The author described the presence of spherocytes (globular-shaped erythrocytes with uniform distribution of hemoglobin) typically found in the blood of patients with hemolytic anemia. Rectal temperatures varied from normal to 40.2°C (104.4°F).

The non-allergenic toxic manifestations of bee venom may be intense and occur so rapidly that it may be difficult to differentiate toxicity from an anaphylactic response. A dog had swallowed a bee (species unknown) and within minutes began vomiting, retching, and evacuating the bowels.[12] Within 5 min the dog began to stagger, drag the hind limbs, and ultimately collapsed, whereupon the owner rushed the pet to a veterinarian. The dog was treated with intravenous antihistamines and seemed to improve sufficiently that it was sent home on oral antihistamine medication. The dog died approximately 10 h later.

The owner of this dog volunteered information that it had been stung by a bee 2 years previously and that the owner had removed the stinger. This seemed to corroborate a diagnosis of anaphylaxis; however, the syndrome did not suggest anaphylaxis nor was the treatment directed at anaphylaxis. It is likely that this was a case of venom toxicity and not an allergic response. To this author's knowledge, there have been no verified cases of dog or cat anaphylaxis from a bee or wasp sting.

Bee venom is known to stimulate contraction of bronchiolar muscle in the cat.[38] There is a significant drop in blood pressure and derangement of the electrocardiogram in both dogs and cats when bee venom is administered intravenously.[39] When a bee sting occurs in a highly vascular area such as a lip or oral mucous membrane, the probability of a direct intravenous injection is high.

A recent report on severe systemic reactions to bee and hornet stings brings insight to the clinical manifestations of bee stings in dogs.[40] One dog was unconscious when presented, with a rectal temperature of 40.5°C, pulse rate of 240 beats per min, and rapid and shallow respirations. The owner reported that the dog had been normal 1.5 h previously. Two hornets (species unknown) were found in the hair coat of the dog and the owner had sprayed a hornet nest earlier in the day. Hematologic findings included leukocytosis (27,000/μl) and a left shift. Serum urea nitrogen (SUN) was elevated (36.6 mg/dl), as was alanine transaminase (ALT) activity (3840 IU/l).

More than 60 honey bee stingers were removed from the oral cavity and ears of a second dog. This dog was ataxic and cyanotic, with a normal body temperature, but a pulse rate of 160 beats per min; 3 h later, the dog began to suffer seizures. The following day, the dog voided bloody urine and vomited blood. Hematologic findings included leukocytosis and a left shift, along with a SUN of 130.4 mg/dl and creatinine of 7 mg/dl. In spite of intensive supportive therapy, the dog died 2 d later with what appeared to be acute disseminated intravascular coagulation (DIC).

A third dog had received only a single sting, but was rushed to an emergency clinic where it was treated with antihistamines and corticosteroids and discharged; 3 d later, the dog was recumbent and passing bloody feces, and 24 h later the dog became icteric and began vomiting a dark brown fluid. This dog responded to intensive supportive therapy.

The clinical signs to note in these cases are prostration, convulsions, bloody diarrhea, vomiting of bloody fluid, leukocytosis with a left shift, and elevated SUN and ALT indicating renal and hepatic involvement. Other studies have demonstrated that hepatic damage also occurs in cats.[41]

C. WILD ANIMALS

Bee and wasp stings occur in wild species, but the prevalence and severity of the stings are unknown. As animals vary in susceptibility to bee venom, it is likely that wild animals cope with bee and wasp venoms as they do other toxic substances in their environment; that is, by avoidance and, perhaps, by increased genetic resistance to stings of species that have evolved with them.

A privately owned 17-year-old tortoise, *Testudo marginata,* was presented with anorexia of sudden onset. On examination, a foreign body was observed imbedded in the soft palate of the mouth. The object was removed and found to be a bee stinger. Corticosteroids and antibiotics were administered and within 36 h the tortoise was eating normally.[42]

Mortality of two individuals in a captive herd of Chacoan peccaries, *Catagonus wagneri,* was originally diagnosed as snakebite, based on the presence of swollen heads and apparent fang marks. A third animal developed facial swelling and depression a few months later. A more thorough search of the enclosure resulted in the location of a wasp nest, *Agelaia* sp., in the sleeping quarters (an earthen tunnel). The carcass of a peccary previously reported missing was found near the nest.[15]

V. TREATMENT

A. FIRST AID

If an owner telephones to report that a pet has been stung by an insect, it is a great help if they can identify the insect. If not, general instructions may be given. It may be impossible to tell where a pet has been stung until a vesicle or inflammatory weal develops. Bees are likely to sting in areas of short hair or that are free of hair. As soon as the injection site has been identified, the area should be inspected for the remains of the stinger if the insect was a honey bee. A retained stinger should be removed by scraping from the side rather than grasping it with tweezers or forceps, which may force additional venom into the victim.[10] Application of a cold compress may diminish initial pain and the degree of local swelling. The pet should be observed for signs of systemic reaction and preparations made for immediate transport to a hospital if signs appear.

B. HOSPITAL TREATMENT

In patients suspected of having a local reaction only, the swollen area should be examined for the presence of a stinger which should be removed as described under first aid. Other possible causes of local inflammation should be differentiated. The administration of antibiotics is a matter of personal choice. Antihistamines and corticosteroids have questionable value once the lesion has developed, but are not contraindicated. Although initial cold packs are indicated to slow or prevent maximum development of a local reaction, once formed, the swelling is more logically treated with hot packs to facilitate increased circulation and absorption of inflammatory products.

Systemic reaction must be treated rapidly and intensively. In human medicine, epinephrine would be administered immediately because most of the peracute systemic reactions are allergic responses and anaphylaxis is the great concern. True allergic reaction to *Hymenoptera* envenomation in animals has not been sufficiently documented to warrant recommendation of routine administration of epinephrine. If signs of respiratory distress, urticaria, diarrhea, vomiting, or salivation are present, a veterinarian should consider administering epinephrine. However, if the animal is in shock from the toxic effects of the venom, epinephrine will be of less value and may be contraindicated. The dose of epinephrine for a dog or cat is 1 to 5 ml of a 1:10,000 aqueous solution administered intravenously, or, if a vein is not accessible, subcutaneously.

The key to intensive therapy is the administration of intravenous multielectrolyte fluids (100 ml/kg/d in a dog, 50 to 75 ml/kg/d in a horse).[40] Corticosteroids and antihistamines are frequently added to the regimen along with medications to alleviate specific signs (i.e., diazepam for convulsions). Careful monitoring of renal and hepatic function is crucial to assessing the progress of a patient. Oxygen supplementation should be available for any animal experiencing respiratory difficulty.

VI. PROGNOSIS

Single stings are rarely fatal. Multiple stings in any animal may be serious and result in a peracute syndrome and death within minutes to hours. The outcome will be dependent on the number of stings inflicted, the species of the bee or wasp, and the age, immune status, and size of the victim.

VII. PREVENTION

It is easy to say that animals should not be placed in high-risk situations. From a practical standpoint, one may suggest only that owners be attentive to hazards that exist in the environment (e.g., wasp nests and beehives).

Desensitization of humans to the allergens in bee venom is currently being conducted on highly sensitive individuals. Studies conducted on laboratory animals indicate that desensitization is effective.[43] However, since the allergic response in animals is as yet unexplored, the technique is not currently useful in veterinary medicine.

REFERENCES

1. **Parrish, H. M.,** Analysis of 460 fatalities from venomous animals in the United States, *Am. J. Med. Sci.,* 245, 129, 1963.
2. **Akre, R. D. and Reed, H. C.,** Biology and distribution of social hymenoptera, in *Handbook of Natural Toxins, Vol. 2. Insect Poisons, Allergens and Other Invertebrate Venoms,* Tu, A. T., Ed., Marcel Dekker, New York, 1984, 3.
3. **Markl, H.,** Superfamily *Apoidea,* (bees), in *Grzimek's Animal Life Encyclopedia,* Vol. 2, Grzimek, B., Ed., Van Nostrand Reinhold, New York, 1975, 454.
4. **Rathmayer, W.,** Hymenoptera: the aculeate wasps, ants and bees, in *Grzimek's Animal Life Encyclopedia,* Vol. 2., Grzimek, B., Ed., Van Nostrand Reinhold, New York, 1975, 421.
5. **Tew, J.,** Honey bees and beekeepers, *Fact Sheet for the Consumer,* United States Department of Agriculture, Washington D.C., 1979.
6. **Harwood, R. F. and James, M. T.,** Stinging insects, in *Entomology in Human and Animal Health,* 7th ed., Harwood, R. F. and James, M. T., Eds., Macmillan, New York, 1979, 422.
7. **Croft, L. R.,** *Allergy to Bee Stings and Its Prevention,* Elmwood Books, Chorley, Lancashire, England, 1988.
8. **van Marle, J.,** Morphology of the venom apparatus, in *Venoms of the Hymenoptera,* Piek, T., Ed., Academic Press, Orlando, FL, 1986, 17.
9. **Maschwitz, U. W. J. and Kloft, W.,** Morphology and function of the venom apparatus of insects — bees, wasps, ants and caterpillars, in *Venomous Animals and Their Venoms,* Vol. 3, Bucherl, W. and Buckley, E. E., Eds., Academic Press, Orlando, FL, 1971, 1.
10. **Caras, R.,** Insects, in *Venomous Animals of the World,* Caras, R., Ed., Prentice-Hall, Englewood Cliffs, NJ, 1974, 88.
11. **Fowler, M. E.,** Insects, in *Current Veterinary Therapy, Food Animal Practice 2,* Howard, J. L., Ed., W. B. Saunders, Philadelphia, 1986, 464.

12. **Antin, I. P.,** Fatal anaphylactic reaction of dog to bee sting, *J. Am. Vet. Med. Assoc.*, 142, 775, 1963.
13. **Breed, M. D.,** Defensive behavior, in *The African Honey Bee,* Spivak, M., Fletcher, D. J. C., and Breed, M. D., Eds., Westview Press, Boulder, CO, 1991, 299.
14. **Eberling, W.,** Wasps, bees and hornets, in *Urban Entomology,* Eberling, W., Ed., University of California, Division of Agricultural Sciences, Los Angeles, CA, 1975, 353.
15. **Benirschke, K., Byrd, M. L., and Meritt, D. A.,** New observations on the Chacoan peccary, *Verhandlungsbericht Erkrankungen Zootiere,* 32, 341, 1990.
16. **Schmidt, J. O.,** Allergy to Hymenoptera venoms, in *Venoms of the Hymenoptera,* Piek, T., Ed., Academic Press, Orlando, FL, 1986, 509.
17. **Hoffman, D. R.,** Insect venom allergy, immunology, and immunotherapy, in *Handbook of Natural Toxins, Vol. 2. Insect Poisons, Allergens and Other Invertebrate Venoms,* Tu, A. T., Ed., Marcel Dekker, New York, 1984, 187.
18. **Shipolini, R. A.,** Biochemistry of bee venom, in *Handbook of Natural Toxins, Vol. 2. Insect Poisons, Allergens and Other Invertebrate Venoms,* Tu, A. T., Ed., Marcel Dekker, New York, 1984, 49.
19. **Piek, T.,** Pharmacology of Hymenoptera venoms, in *Handbook of Natural Toxins, Vol. 2. Insect Poisons, Allergens and Other Invertebrate Venoms,* Tu, A. T., Ed., Marcel Dekker, New York, 1984, 135.
20. **Banks, B. E. C. and Shipolini, R. A.,** Chemistry and pharmacology of honey-bee venom, in *Venoms of the Hymenoptera,* Piek, T., Ed., Academic Press, Orlando, FL, 1986, 330.
21. **Habermann, E.,** Chemistry, pharmacology and toxicology of bee, wasp and hornet venoms, in *Venomous Animals and Their Venoms,* Vol. 3, Bucherl, W. and Buckley, E. E., Eds., Academic Press, Orlando, FL, 1971, 61.
22. **Nakajima, T.,** Pharmacological biochemistry of vespid venoms, in *Venoms of the Hymenoptera,* Piek, T., Ed., Academic Press, Orlando, FL, 1986, 309.
23. **Piek, T.,** Historical introduction, in *Venoms of the Hymenoptera,* Piek, T., Ed., Academic Press, Orlando, FL, 1986, 1.
24. **Piek, T.,** Venoms of bumble-bees and carpenter-bees, in *Venoms of the Hymenoptera,* Piek, T., Ed., Academic Press, Orlando, FL, 1986, 417.
25. **Piek, T. and Spanjer, W.,** Chemistry and pharmacology of solitary wasp venoms, in *Venoms of the Hymenoptera,* Piek, T., Ed., Academic Press, Orlando, FL, 1986, 161.
26. **Halliwell, R. E. W. and Gorman, N. T.,** Nonatopic allergic skin diseases, in *Veterinary Clinical Immunology,* Halliwell, R. E. W. and Gorman, N. T., Eds., W. B. Saunders, Philadelphia, 1989, 270.
27. **Morrow, A. N., Quinn, P. J., and Baker, K. P.,** Some characteristics of the antibodies involved in allergic skin reactions of the horse to biting insects, *Br. Vet. J.*, 143, 59, 1987.
28. **Blood, D. C. and Radostits, O. M.,** *Veterinary Medicine,* 7th ed., Bailliere Tindall, London, 1989, 1357.
29. **Pascoe, R. R.,** Equine Dermatoses, in *University of Sydney, The Post-Graduate Foundation in Vet. Sci.,* Veterinary Review No. 22, Sydney, Australia, 1981, 60.
30. **Pascoe, R. R.,** *Colour Atlas of Equine Dermatology,* Wolfe Publishing, London, 1989, 69.
31. **Williams, D. J., Stuart, B. P., and Gaskin, C. B.,** Insect stings in swine, *Vet. Med. Small Animal Clinician,* 71, 1663, 1979.
32. **Meerdink, G. L.,** Bites and stings of venomous animals, in *Current Veterinary Therapy, VII. Small Animal Practice,* Kirk, R. W., Ed., W. B. Saunders, Philadelphia, 1980, 174.
33. **Meerdink, G. L.,** Bites and stings of venomous animals, in *Current Veterinary Therapy, VIII. Small Animal Practice,* Kirk, R. W., Ed., W. B. Saunders, Philadelphia, 1983, 155.
34. **Mount, M. E.,** Biotoxins, in *Textbook of Veterinary Internal Medicine,* 3rd ed., Ettinger, S. J., Ed., W. B. Saunders, Philadelphia, 1989, 472.
35. **Kral, F. and Novak, B. J.,** *Veterinary Dermatology,* Lippincott, Philadelphia, 1953, 308.
36. **Wysoke, J. M., Van-den Berg, F., and Marshall, C.,** Bees sting-induced haemolysis, spherocytosis and neural dysfunction in three dogs, *J. South Afr. Vet. Assoc.*, 61, 29, 1990.

37. **Muller, G. H., Kirk, R. W., and Scott, D. W.,** Hymenoptera, in *Small Animal Dermatology,* 4th ed., Muller, G. H., Kirk, R. W., and Scott, D. W., Eds., W. B. Saunders, Philadelphia, 1989, 422.
38. **Chand, N. and DeRoth, L.,** *In vitro* peripheral airway constriction to honey bee venom in the cat, *Vet. Human Toxicol.,* 22, 244, 1980.
39. **Kaplinsky, E., Ishay, J., Ben-Shachar, D., and Gitter, S.,** Effects of bee *Apis mellifera* venom on the electrocardiogram and blood pressure, *Toxicon,* 15, 251, 1977.
40. **Cowell, A. K., Cowell, R. L., Tyler, R. D., and Nieves, M. A.,** Severe systemic reactions to Hymenoptera stings in three dogs, *J. Am. Vet. Med. Assoc.,* 198, 1014, 1991.
41. **Neuman, M. B., Ishay, J. S., and Eshchar, J.,** Hornet *Vespa orientalis* venom sac extract causes hepatic injury in cats, *Comp. Biochem. Physiol.,* 74, 469, 1983.
42. **Berry, A. P.,** Inappetence in a tortoise, *Vet. Rec.,* 1211D, 207, 1987.
43. **Rothenbacher, H. and Benton, A. W.,** Pathologic features in mice hyposensitized to bee venom, *Am. J. Vet. Res.,* 33, 1867, 1972.

Chapter 6

BLISTER BEETLE POISONING

I. ETIOLOGY

The 290,000 named species of beetles distributed worldwide comprise 40% of named insects. Blister beetles, *Epicauta* spp., are in the family *Meloidae*, order *Choleoptera,* class *Insecta.* Numerous species of the genus have been implicated in poisoning of horses and livestock in North America.[1-5]

Beetles may be beneficial to human beings (ladybugs, family *Coccinellidae)* or be extremely destructive (cotton boll weevil, *Anthonomus grandis).* Some 3000 species of beetles in the family *Meloidae* are called "blister beetles" because of variable amounts of an irritant chemical (cantharidin, which inhibits predation)[4,6] contained in the hemolymph and gonads.

Although 200 species of blister beetles are found in the U.S., mammals have been poisoned by accidental ingestion of feed contaminated with the dead bodies of only a few species (see Table 1). However, surveys of blister beetles for cantharidin indicate that many more species are potentially toxic if specific environmental conditions exist.[7]

The preferred diet of adult *Epicauta* spp. involved with poisoning consists of nectar and pollen of flowering plants along with small amounts of foliage. Eggs are laid and hatch in the soil. The larvae of known poisonous species of *Epicauta* feed on grasshopper eggs, but some other species feed on the eggs of solitary ground nesting bees.[8]

A. GEOGRAPHICAL DISTRIBUTION

Although blister beetles are found in virtually every state in the U.S., poisoning has been reported only in Arizona,[3] Colorado,[2] New Mexico,[9] Illinois,[9] Florida,[1,9] Oklahoma,[4,5,10,11] Texas,[3] and Tennessee.[12] Poisoning has also been documented in horses in northern Mexico and in cattle in South Africa.[3]

B. HOSTS

Horses are most often poisoned by ingestion of dead beetles in the hay, perhaps because their simple stomach allows for more rapid absorption of the poison.[1,4,9,12-22] However, the following hosts have also been poisoned in natural or experimental situations: cattle,[23] sheep,[13,23] goats,[13] rabbits,[13] rats,[13] hedgehogs,[24] dogs, and emus.[25-27] Poisoning in human beings has been noted for decades and is usually the result of either improper medical use of cantharides or malicious poisoning.[28]

C. ENVIRONMENTAL CONDITIONS

Poisoning has been primarily restricted to horses consuming alfalfa hay, *Medicago sativa.* Hays other than alfalfa have rarely been incriminated, except

TABLE 1
Species of Blister Beetle Reported to be the Cause of Poisoning in Horses and Livestock

Species	Distribution	Ref.
Epicauta occidentalis[a]	Nebraska, Kansas, Oklahoma, Texas, Louisiana	1, 4, 5, 13
E. lemniscata[a]	Same as above plus east to New Jersey and south to Florida	1, 4, 5, 13
E. vittata	Iowa east to Maine and New Jersey	4, 5, 13
E. pardalis	Southwestern U.S. (Colorado) and northwestern Mexico	2, 3
E. pennsylvanica	Canada to Mexico, Utah to Atlantic coast	9
E. temexa	Texas	3
E. albida	Midwest U.S. and Southwest (Texas)	3
E. attrivittata	Same as above	3
Lytta vesicatoria, Spanish fly	South Africa, cattle	13

[a] Most common species involved.

when significant quantities of flowering weeds are present to provide forage for the beetles.

A major change in methods of harvesting alfalfa beginning in the mid-1960s provided conditions conducive to the inclusion of blister beetles in the hay in sufficient quantity to cause lethal poisoning if ingested.[29] Blister beetles feed on the flowers of alfalfa and may congregate in swarms that move about a field in a random manner. In the past, mowing, drying, and baling techniques allowed the beetles to move out of the drying hay. Now, hay is cut, crimped, and swathed in a single operation to facilitate quick drying. A swarm of beetles may be crushed and left to dry in a section of the hay, which is then incorporated into the bale, perhaps even into a single flake.[29]

Other conditions needed for poisoning are the presence of grasshopper eggs upon which the beetle larvae feed and a beetle species that is colonial and swarming, thus present in sufficient numbers to pose a risk. Beetles may be observed swarming in front of the windrower.[3]

Poisoning may be confined to a single animal in a herd or multiple animals may be involved.

II. POISON

Cantharidin is the most important active principle in cantharides, which historically was obtained from dried Spanish flies (blister bugs), *Lytta*

(Cantharis) vesicatoria. Medically, cantharides were applied externally as a blistering agent and also administered orally in minute quantities as a diuretic and aphrodisiac.[28,30] Purified, crystalline cantharidin is a potent vesicating agent that is readily absorbed through the gastrointestinal mucosa and, to a lesser degree, through the skin.[4] As little as 4 g of dried beetles may be lethal to a horse. One investigator collected up to 145 g dried beetles from a single flake of alfalfa hay weighing 2 kg.[11]

The concentration of cantharidin present in a beetle varies from 1 to 5% of the dried weight of the beetle.[3,23] Males always have a higher concentration because the male produces the cantharidin and transfers a quantity to the female at copulation.[29] The oral experimental lethal dose of crystalline cantharidin for dogs and cats was 1.0 to 1.5 mg/kg and 20 mg/kg for rabbits.[13] It is estimated that the lethal dose for a human being is less than 1.0 mg/kg and for the horse approximately 0.5 mg/kg.[13]

III. CLINICAL SIGNS

Cantharidin produces an intense, direct irritant effect on the skin and mucous membranes of the esophagus, stomach, and intestines. Once absorbed, the toxin may affect many different organs. Excretion is via the kidneys, resulting in transfer of the irritant effect to the urinary tract, particularly the bladder and urethra.[4,5]

Clinical signs, as described in the horse, vary with the dose ingested. Massive doses may cause shock and death within 4 h.[1,5] Smaller doses cause gastroenteritis, nephrosis, cystitis, and urethritis; thus, signs may include anorexia, soft feces or mucoid to bloody mucous on the feces, intestinal atony, colic, dysuria (frequent, painful urination, or oliguria to anuria), and hematuria. Diarrhea may be observed in animals that live for a few days. The body temperature may elevate to 41.1°C (106°F). Other signs observed include depression, weakness, muscle rigidity, collapse, prostration, dehydration, and sweating.[1,9-12,15,31]

Animals frequently become dyspneic (tachypnea) and rales may be heard on auscultation due to pulmonary edema. Myocarditis may initiate cardiovascular signs including tachycardia, congested mucous membranes, slow capillary refill time, and synchronous diaphragmatic flutter (probably caused by the hypocalcemia).[4] Ulceration of the oral mucosa may be observed and a horse may be seen to splash its muzzle into water without drinking.[4]

The course of the disease may be as short as 4 h, with massive dose ingestion, to 5 d in lethal poisoning. Approximately 50% of affected horses have died.[1] If a horse lives for more than 1 week, the prognosis becomes favorable.

IV. DIAGNOSIS

A combination of clinical signs, alteration in hematologic and serum chemistry values, lesions, and analysis for cantharidin provide a definitive diagnosis.

Hay and intestinal contents of dead animals should be inspected for evidence of the beetles, but the offending pocket of beetles in a bale of hay may be gone.

A. CLINICAL PATHOLOGY

The reported data refer to poisoning in the horse. In both field cases and experimental poisoning, there is an elevated serum protein and packed cell volume (59%) caused by dehydration and shock. The damaged gastrointestinal mucosa allows rapid invasion of enteric bacteria, causing a bacteremia and leukocytosis (17,000/mm^3). There is mild elevation in serum urea nitrogen (50 mg/dl), hypocalcemia (5.9 mg/dl, normal 12.8 ± 1.2),[4,27] and hypomagnesemia (0.7 to 1.8 mg/dl, normal 2.5 ± 0.3).[27] Specific gravity of urine is low in the early stages of the disease (1.003 to 1.006) and erythrocytes are usually present in the urine, yielding a positive occult blood reaction.

B. NECROPSY

Gross lesions may be minimal with massive dose ingestion.[4,22] In more protracted cases, oral ulcers, vesication, and desquamation of patches of the distal esophagus, erosions and ulceration of the gastrointestinal tract, mucus in the renal pelvis, and renal cortical hemorrhages may be seen. Hyperemia and hemorrhages are seen in the urethra and bladder mucosa. Ventricular myocarditis, pulmonary edema, petechial hemorrhages of serosal surfaces, hepatomegaly, and splenomegaly[5] are other signs.

The initial microscopic lesion is acantholysis of the mucosa of the gastrointestinal tract, epithelium of the urinary tract, and endothelium of vessels.[32] Other microscopic lesions include myocarditis, renal tubular nephrosis, and degenerative changes in the kidneys and digestive tract.[4] Details of the pathology are fully described in the literature.[4,33]

C. DIFFERENTIAL DIAGNOSIS

In the horse, other diseases that must be considered include primary diseases of the digestive tract producing colic, castor bean *(Ricinus communis)* poisoning, monensin poisoning, *Escherichia coli* endotoxemia, colitis X, and any disease producing shock.

D. ANALYTICAL

Cantharidin may be detected in urine, tissue (kidney and blood), gastrointestinal contents, and the dried beetles themselves by high pressure liquid chromatography (HPLC),[13,14,24] or gas chromatography/mass spectrometry (GC/MS).[3] Cantharidin is excreted rapidly and may not be present in detectable amounts after 4 to 5 d following ingestion.[24]

The key factors in a differential diagnosis of blister beetle poisoning in a horse are: finding the beetles in hay being fed, clinical signs of colic and hematuria, an elevated packed cell volume, and hypocalcemia and hypomagnesemia. The diagnosis may be confirmed by analysis for cantharidin, or the finding of beetle parts in the ingesta.

V. TREATMENT

There is no specific treatment. The administration of either activated charcoal or mineral oil (but not together) via a gastric tube may aid animals that have consumed a small dose or are in the early stages of poisoning. General supportive therapy should include correction of fluid loss and electrolyte imbalance, particularly hypocalcemia and hypomagnesemia.[11,33]

Broad-spectrum antimicrobial therapy is required to counter secondary bacterial invasion from the gastrointestinal tract. Aminoglycoside antibiotics and others that are potentially nephrotoxic or that are excreted via the kidney should be avoided, as nephrosis is an effect of cantharidin.

VI. PROGNOSIS

The prognosis depends on the degree of irritation to parenchymal organs (heart, liver, and kidney) and the gastrointestinal and urinary tracts, which in turn is correlated with the ingested dose. The prognosis is grave for animals with signs of severe shock and hematuria.

VII. PREVENTION OF POISONING

In areas where blister beetle poisoning is known to occur, hay should be inspected by the owner or the caretaker as the hay is fed.[29,33] Unfortunately, because alfalfa hay may be transported thousands of miles before ultimate consumption, one cannot depend on the absence of swarming blister beetles in the local area. Alfalfa will not be attractive to blister beetles unless well into the flowering or bloom stage.[8] The person driving the windrower/crimper may minimize the problem. For example, if beetles are observed swarming in front of the cutting bar, the operator may stop to allow the beetles to disperse or move to another area of the field to continue harvesting.[8] Since the presence of most species of blister beetles is correlated with grasshoppers, a regular program to control them is beneficial. Insecticide application to kill the beetles would be a last resort.[8] Consultation with a local County Agricultural Extension agent should be sought.

REFERENCES

1. **MacKay, R. J. and Wollenman, P.,** An outbreak of blister beetle poisoning in horses in Florida, *Florida Vet. J.*, 10, 11, 1981.
2. **Kruzel, M. K.,** More beetles to bug us, *Equus*, 56, 91, 1982.
3. **Ray, A. C., Kyle, A. L. G., Murphy, M. J., and Reagor, J. C.,** Etiologic agents, incidence, and improved diagnostic methods of cantharidin toxicosis in horses, *Am. J. Vet. Res.*, 50, 187, 1989.

4. **Schoeb, T. R. and Panciera, R. J.**, Blister beetle poisoning in horses, *J. Am. Vet. Med. Assoc.*, 173, 75, 1978.
5. **Schoeb, T. R. and Panciera, R. J.**, Pathology of blister beetle *Epicauta* poisoning in horses, *Vet. Pathol.*, 16, 18, 1979.
6. **Carrel, J. E. and Eisner, T.**, Cantharidin: potent feeding deterrent to insects, *Science*, 183, 755, 1974.
7. **Capinera, J. L., Gardner, D. R., and Stermitz, F. R.**, Cantharidin levels in blister beetles, *(Coleoptera: Meloidae)* associated with alfalfa in Colorado, *J. Economic Entomol.* 78, 1052, 1985.
8. **Hutchison, W. D., Murphy, M. J., and Tufte, G. N.**, Blister Beetles in Alfalfa, Management Options to Minimize Poisoning in Horses, AG- FO 5510-D, Minnesota Agricultural Extension Service, University of Minnesota, St. Paul, 1990.
9. **Beasley, V. R., Wolf, G. A., Fisher, D. C., et al.**, Cantharidin toxicosis in horses, *J. Am. Vet. Med. Assoc.*, 182, 283, 1983.
10. **Panciera, R. J.**, Cantharidin (blister beetle) poisoning, in *Equine Medicine and Surgery*, 2nd ed., Catcott, E. J. and Smithcors, J. F., Eds., American Veterinary Publications, Wheaton, IL, 1972, 30.
11. **Panciera, R. J.**, Cantharidin (blister beetle) poisoning, in *Equine Medicine and Surgery*, 3rd ed., Vol. 1, Mansmann, R. A. and McAllister, E. S., Eds., American Veterinary Publications, Santa Barbara, CA, 1982, 203.
12. **Bahme, A. J.**, Cantharides toxicosis in the equine, *Southwestern Vet.*, 21, 147, 1968.
13. **Ray, A. C., Tamulinas, S. H., and Reagor, J. C.**, High pressure liquid chromatographic determination of cantharidin, using a derivatization method in specimens from animals acutely poisoned by ingestion of blister beetles, *Epicauta lemniscata, Am. J. Vet. Res.*, 40, 498, 1979.
14. **Ray, A. C., Post, L. O., and Reagor, J. C.**, Confirmation of cantharidin toxicosis due to ingestion of blister beetles, *Vet. Human Toxicol.*, 22, 398, 1980.
15. **Fowler, M. E.**, Blister beetles, in *Current Veterinary Therapy, 3. Food Animal Practice*, Howard, J. L., Ed., W. B. Saunders, Philadelphia, 1992, in press.
16. **Moore, R. W.**, Cantharides poisoning in a horse, *Vet. Med.*, 58, 961, 1963.
17. **Lane, T. J.**, Blister beetle poisoning in horses, *Florida Horse*, 25, 1846, 1983.
18. **Lane, T. J.**, Blister beetle poisoning in horses, *Equine Vet. Data*, 4, 360, 1983.
19. **Rollins, J. B.**, Blister beetle poisoning in horses, *Equine Practitioner*, 7, 6, 1985.
20. **Schmitz, D. G. and Reagor, J. C.**, Cantharidin (blister beetle) toxicity, in *Current Therapy in Equine Medicine*, Robinson, N. E., Ed., W. B. Saunders, Philadelphia, 1987, 120.
21. **Schmitz, D. G.**, Cantharidin toxicosis update, *Equine Vet. Data*, 9, 87, 1988.
22. **Butler, L. and Kitchen, D.**, A case of blister beetle poisoning in a horse, *Southwestern Vet.*, 38, 13, 1987.
23. **Sippel, W. L.**, Blister beetle poisoning, *Cattleman*, 63, 208, 1976.
24. **Ray, A. C., Post, L. O., Hurst, J. M., Edwards, W. C., and Reagor, J. C.**, Evaluation of an analytical method for the diagnosis of cantharidin toxicosis due to ingestion of blister beetles *Epicauta lemniscata* by horses and sheep, *Am. J. Vet. Res.*, 41, 932, 1980.
25. **Rabkin, S. W., Friesen, J. M., Ferris, J. A. J., et al.**, A model of cardiac arrhythmias and sudden death: cantharidin induced toxic cardiomyopathy, *J. Pharmacol. Exp. Therapy*, 210, 43, 1970.
26. **Opie, E. L.**, Lymph formation and edema of the liver with experimental nephritis produced by cantharidin, *J. Exp. Med.*, 16, 831, 1912.
27. **Pearce, R. M.**, The renal lesion of experimental cantharidin poisoning, *J. Exp. Med.*, 17, 543, 1913.
28. **Shawley, R. V. and Rolf, L. L., Jr.**, Experimental cantharidiasis in the horse, *Am. J. Vet. Res.*, 45, 2261, 1984.
29. **Arnold, D. C.**, Blister beetles of Oklahoma, *Technical Bulletin T-145*, Agricultural Experiment Station, Oklahoma State University, Stillwater, 1976, 19.

30. **Davenport, J.,** Cantharidis, in *Aphrodisiacs and Love Stimulants*, Walton, A. H., Ed., Luxor Press, London, 1865, 40.
31. **Sierra, J. R., Woggon, W.-D., and Schmid, H.,** Transfer of cantharidin during copulation from the adult male to the female *Lytta vesicatoria* (Spanish flies), *Experimentia*, 32, 142, 1976.
32. **Weakley, D. R. and Einbinder, J. M.,** The mechanism of cantharidin acantholysis, *J. Invest. Dermatol.*, 39, 39, 1961.
33. **Oehme, F. W.,** Blister beetle, in *Current Therapy in Equine Medicine*, Robinson, N. E., Ed., W. B. Saunders, Philadelphia, 1983, 588.

Chapter 7

SPIDERS

I. IDENTIFICATION AND BIOLOGY

There are over 100,000 species of spiders. Most of them have a biting apparatus and associated venom glands that produce venom used in obtaining food.[1] Only a few species produce venom of sufficient potency to cause illness in larger vertebrates or have chelicerae (jaws) capable of penetrating vertebrate skin (Table 1).[2] Spiders are beneficial to humankind, yet many people believe that every spider is waiting to envenomate and kill any unsuspecting person happening by.[3]

A. MORPHOLOGY

Spiders have eight pairs of legs (Figure 1). The pedipalp of some species may be large enough to be mistaken for a pair of legs. The pedipalp serves as a tactile organ in both sexes and, in the male of some species, it has been modified to become an intromittent organ for placement of sperm.[1] The body is separated into a cephalothorax cranially and an abdomen caudally (Figure 1). The relative size and shape of these structures vary from species to species.[2,4]

On the rostral end of the cephalothorax are a pair of chelicerae with associated venom glands ducted to the hollow, cornified fangs at the tip (Figure 2).[2,5] The chelicerae are the biting mouth parts of a spider. All spiders are predators, feeding primarily on other invertebrates. Hunting spiders have multiple (usually eight) eyes, with sharp visual acuity.

Some spiders build elaborate webs for catching prey. The thread is produced by spinning organs (spinnerets) on the caudal ventral abdomen.

B. TERMINOLOGY

Venomous spiders may be small or large. Large, hairy spiders are universally feared as being highly venomous, but not all are so.[6] The name *tarantula* causes confusion. It was first applied to a venomous European wolf spider, *Lycosa tarentula* (family, *Lycosidae),* in Italy in the vicinity of Taranto. The inhabitants of Taranto developed arachnophobia to the extent of resorting to a ritual, frenzied dancing (tarantella) to rid the body of venom from real or imagined spider bites.[1] The name confusion is compounded by the fact that the generic name "*Tarantula*" is used for a tailless whip scorpion.[1]

American tarantulas belong to a different family of spiders, *Theraphosidae*. Approximately 40 species live within the confines of the U.S., primarily in the Southwest (Figure 3) with a few species ranging as far as northern California, Oregon, Utah, and southwestern Idaho.[5] None are found east of the Mississippi River. The majority of these large spiders are not highly venomous and represent no threat to humans or animals in spite of their ferocious appearance.[1] However, all American tarantulas have large chelicerae and may produce a

TABLE 1
Classification of Venomous Spiders

Phylum — *Arthropoda*
Class — *Arachnida*
Order — *Araneida (Araneae)* — spiders
Suborder — *Mygalomorphae (Orthognatha)* — tarantulas
  Family — *Dipluidae* — funnel-web spiders
    Trechona spp.
    Atrax robusta — Sydney funnel-web spider
    Ixeuticus robustus — black house spider (Australia)
  Family — *Theraphosidae (Aviculariidae)* — tarantulas or bird spiders
    Lasiodora klugi —
    Aphonopelma spp. — Western tarantula
Suborder — *Araeneomorphae (labidognatha)*
  Family — *Lycosidae* — wolf spiders
    Lycosa spp. —
  Family — *Theridiidae* —
    Latrodectus spp. — black widow spider
    Latrodectus tredecimguttatus — karakurt (black wolf)
  Family — *Loxoscelidae (Scytodidae)*
    Loxosceles spp. — brown, violin, or recluse spiders

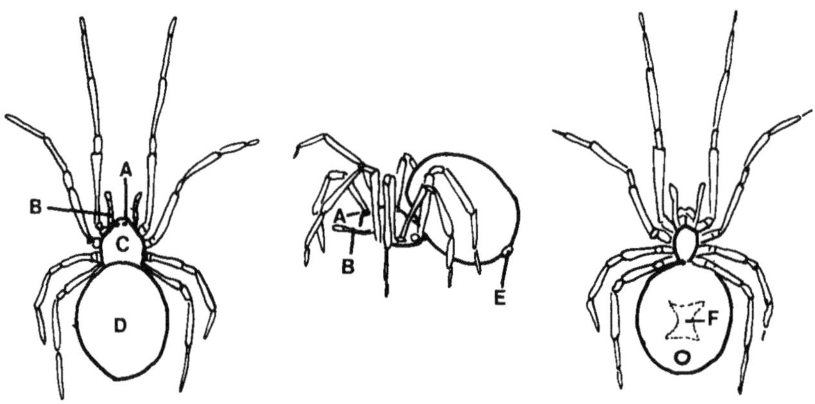

FIGURE 1. Diagram of the basic morphology of a spider, *Latrodectus mactans*. A, eye; B, pedipalp; C, cephalothorax; D, abdomen; E, spinneret; and F, red hourglass mark. (Modified from Wong, R. C., Hughes, S. E., and Voorhees, J. J., Arch. Dermatol., 123, 98, 1987.)

painful bite. The Mexican red- or pink-kneed tarantula, *Aphonopelma* spp., (Figure 4) unfortunately is commonly sold in the pet trade. The less ornate *Aphonopelma californica* is common in California and other areas of the Southwest.

Some large tarantulas (up to 25 cm with legs extended) prey upon snakes and birds. Presumably, these species would be a hazard for animals that harass the spiders. Included in this group are the North and South American species

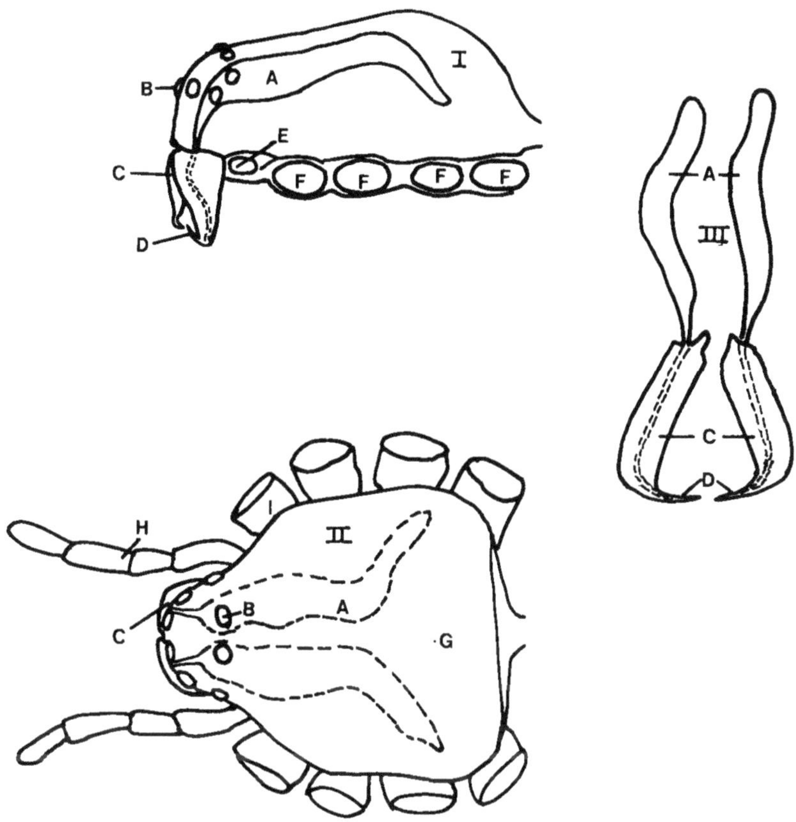

FIGURE 2. Schematic diagram of the venom apparatus of a spider, *Latrodectus mactans*. I, lateral view of the cephalothorax; II, dorsoventral view of the cephalothorax; III, isolated chelicerae and venom glands. A, venom gland; B, eye; C, chelicera; D, fang; E, base of pedipalp; F, base of leg; G, cephalothorax; H, pedipalp. (Modified from Wong, R. C., Hughes, S. E., and Voorhees, J. J., *Arch. Dermatol.*, 123, 98, 1987. With permission.)

of *Aphonopelma, Avicularia, Lasiodora, Techona,* and *Grammostola; Harpactirella* in Africa; and *Atrax* and *Ixeuticus* in Australia.

C. VENOMOUS SPECIES

The prevalence of spider bites in humans is not well understood. Parrish[7] analyzed 460 fatalities caused by venomous animals in the U.S. occurring from 1950 to 1959. He reported 65 spider bite fatalities as compared with 94 rattlesnake bites and 229 hymenoptera stings.

It is impossible to determine the prevalence of spider bites in animals. Neither is it possible to make an authoritative list of spider genera and species involved in animal bites. Spider bite is usually a subjective diagnosis, based on preconceived ideas of the signs that should be produced by a spider bite. Approximately 60 species of spiders in the U.S. have been implicated in

FIGURE 3. Tarantula of the southwestern U.S. *Aphonopelma eutylenum*

FIGURE 4. Mexican red-kneed tarantula, *Brachypelma smithii*.

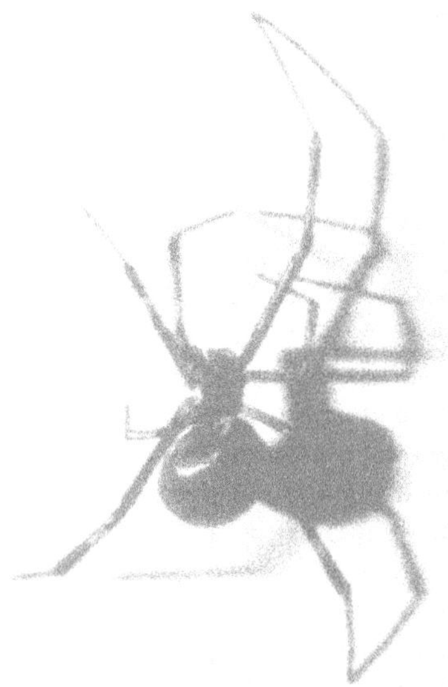

FIGURE 5. Black widow spider, *Latrodecus mactans.*

medically significant bites in humans.[5] Much of the spider bite literature in veterinary medicine is extrapolated directly from human envenomation. Experimental data to verify that animals react to spider bites as do humans is lacking.

The two most important venomous spiders in the U.S. are the widow spiders, *Latrodectus* spp., (family *Theridiidae)* and the recluse spiders, *Loxosceles* spp., (family *Loxoscelidae* or *Scytodidae).* See Wong's review[5] for a listing of the genera and syndromes involved in human envenomation.

Widow spiders, *Latrodectus* spp., are found throughout the temperate and tropical zones of the world.[8] The body is approximately 15 mm long and the legs may spread up to 40 mm. Typically, female widow spiders have a shiny, dark brown or black coloration of the globulous abdomen, with variable-colored markings on either the ventral or dorsal surface. *Latrodectus mactans* (murderer) is the common black widow spider of the southern U.S., ranging into Mexico (Figure 5). This spider has a red hour-glass mark on the ventral abdomen. There is local variation in the configuration of the red mark. An Australian species, *L. hasselti,* has sufficient red on the body to be called the redback spider.[8]

The male *Latrodectus* spp. is much smaller than the female and is inoffensive and not involved in human or animal envenomation. Contrary to legendary belief, the female does not always kill and consume the male after mating.

At least 10 species of recluse spiders *Loxosceles* spp. inhabit the contiguous U.S., only six of which *(L. reclusa, laeta, unicolor, arizonica, devia,* and *rufescens)* have been reported to envenomate humans and cause necrotic arachnidism.[5] The species vary in body size from 10 to 15 mm in length and 4 to 6 mm in width, and may have a leg span of more than 25 mm. A dark brown, violin-shaped marking is usually located on the dorsum of the cephalothorax, giving rise to one of the common names for these spiders (fiddleback or violin spider). There are three pairs of eyes in contrast to the usual four pairs of eyes in other arachnids.[5]

Spiders are frequently transported from one country to another via packing containers or produce, so that bites of non-native species may occur.

D. DISTRIBUTION

Venomous spiders are found worldwide. *Latrodectus* spp. is found in North America, Australia, Africa, and the Mediterranean. *Latrodectus tredecimguttatus* (karakurt) is the common species found in the Mediterranean and southern Asian regions.[9] The brown recluse spider, *Loxosceles reclusa,* is found primarily in the south, from Texas east to Florida, and in the midwest; but specimens may be accidentally transported to other areas in furniture and other household goods being moved about the country. Most reports of bites from *Loxosceles* have involved *L. reclusa.* As the name implies, this species is shy and retiring and is nocturnal. It avoids lighted areas and is found in attics, closets, garages, storage areas for clothing and bedding, around rubbish, and inside overstuffed furniture.[5] One or more species of *Loxosceles* may be found in most regions of the contiguous U.S. and South America.[5] More precise information on the world distribution of venomous spiders is to be found in the literature cited.[2,4]

Atrax spp. are found in many locations and habitats in Australia except the central and western deserts. *Atrax robusta,* the Sydney funnel-web spider, is the most notorious, and is found in the vicinity of Sydney.[10]

II. CONDITIONS OF POISONING

Spiders are generally shy and will withdraw if possible. Animal spider encounters are bound to occur if an animal accidently blunders into a web or harasses a spider. Young animals are more likely to be attracted to diurnal spiders moving about an area, and may even attempt to play with them.

Black widow spiders are more aggressive at the time they are guarding an egg sac. Pets are not likely to place paws into corners, boxes, cupboards, storage areas, or lift wood from a woodpile or investigate a pile of rags as a human would do. Otherwise, pets would be at the same risk of exposure to spider bite as a human child. Pets may encounter spiders in a basement, closet, or attic. Many of the *Latrodectus* spp. do not live in or near human habitation, but are found in fields, building webs in crevices in the soil or in vegetation. Human poisoning occurs most commonly in workers that are hand-sheathing shocks of grain.[9]

Spider bite in livestock and horses is rarely reported. An exception is the case of the black house spider, *Ixeuticus robustus,* in Australia. This inland Queensland spider is nocturnal and frequently seen on the floor of stables at night.[11] Bites are usually seen on the head and neck, presumably incurred while the horse sleeps.[11]

Spider bite has never been reported in free-ranging wild animals. Captive wild animals are at the same risk as domestic animals. Only the females of *Latrodectus* spp. bite, while both male and female *Atrax maculatus* bite.

III. VENOM

Spider venom is complex and varies from species to species. In addition to numerous enzymes, other nonprotein substances (γ-aminobutyric acid, polypeptides) may contribute to the toxic effects.[5,12,13]

A. *LATRODECTUS* VENOM

One of the lethal fractions is a labile protein,[14] primarily a neurotoxin. The toxic protein causes release of neurotransmitters, like noradrenaline and acetylcholine, ultimately producing an ascending motor paralysis and destruction of peripheral nerve endings.[8]

Toxicity of a spider venom depends upon the age and species of spider (older individuals are more venomous), location of the bite, the amount injected by the spider, age and condition of the victim, and season of the year. It is estimated that drop for drop, the venom of *Latrodectus mactans* is 15 times more toxic than *Crotalus* rattlesnake venom.[12,15]

In addition to venom, urticarial-inducing hairs are a second defense mechanism of tarantulas.[16] The hairs are present on the abdomen and limbs. A threatened spider may rub its abdomen with a leg to dislodge the hair toward an enemy.

IV. CLINICAL SIGNS

Most of the signs attributed to spider bites in animals seem to be extrapolated from reports of human envenomation.[8] There are few reports of experimental envenomation studies to delineate the precise progression of envenomation in animals. Limited studies on laboratory animals show variation in effects. Anaphylaxis does not appear to be part of the syndrome of spider bite in animals and is rare in human envenomation.

A. *LATRODECTUS* ENVENOMATION

In humans, a transient pin-prick pain is associated with the bite. Tiny red marks and slight erythema may be observed on light human skin, but is easily overlooked in animal bites. If envenomation is severe, muscle fasciculation begins within minutes (10 to 60), leading to muscle rigidity and abdominal pain, numbness spreading from the bite site, ataxia, excessive salivation, tonic clonic convulsion, and finally, flaccid paralysis. The excessive muscular activity causes an elevation in body temperature. Other clinical signs reported in humans include headache, dizziness, weakness, fever, salivation, nausea, vomiting, slurring of speech, increase in blood and cerebrospinal fluid pressure, backache, and dyspnea.[17] Mild forms of black widow spider bite are known to occur.

Some of the foregoing signs, in varying degrees, may be observed in animals. In dogs, envenomation results in hyperesthesia accompanied by progressive, dull muscular pain, hypertension, muscle fasiculation, and intense excitability.[18-21] Abdominal muscle rigidity is a classic sign, and tonic-clonic convulsions may occur.[19]

Clarke[22] reports that bites in dogs are frequent and fatalities have occurred. The bite is characterized by edema and pain at the bite site, emesis, and rigidity of the abdominal muscles. There may be weakness, dyspnea, anorexia, and progression toward paralysis. Death may occur in as short a time as 4 h or as long as several days.[22]

Cats are extremely sensitive to the toxin, and severe paralytic signs may be noted early. Additional signs observed in cats include salivation, diarrhea, and vomiting. Afflicted cats may lose up to 20% of their body weight in the first 24 h following envenomation.[19]

Camels are thought to be highly sensitive to the bites of *Latrodectus tredecimguttatus*, but no signs or lesions have been reported.[9,23-25] Losses of horses, cows, sheep, cats, dogs, and camels by *Latrodectus* bites have been reported in Europe, South America, and India.[26-29]

B. *LOXOSCELES RECLUSA* ENVENOMATION

Approximately 12 species of *Loxosceles* are native to the U.S. Only two or three cause necrotic arachnidism in humans. Animal bites are rarely reported. The bite is relatively painless and may not be noticed. The progression of signs of severe envenomation occuring in humans is as follows:[5,30] a blue-gray macular halo develops around the puncture site in a few hours to a few days (hemolysis and arterial spasm producing cyanosis). The victim may experience pain at the bite site, a tingling sensation, and pruritis. A cyanotic, irregularly shaped pustule/vesicle/bulla may appear at the bite site. That lesion is surrounded by erythema and edema that may spread peripherally to engulf a limb or the torso. Ultimately, the vesicle or bulla ruptures, leaving a necrotic ulcer that may require as long as 6 months to heal, leaving an unsightly scar.[5] Systemic manifestations may include generalized urticaria, malaise, icterus, headache, convulsions, hemolysis, disseminated intravascular coagulation, fever, vomiting, diarrhea, hemoglobinuria, hematuria, anuaria, delirium, shock, and coma.[5]

In dogs and cats, systemic reactions are rare, but may include hemolysis, thrombocytopenia, fever, weakness, joint pain, and possibly death.[19] The dermal lesion is an indolent ulcer that may take several days to become evident and several months to heal.[19,31]

Berger[32] is of the opinion that many bites by *Loxosceles* are mild and never reach the necrotic ulceration stage.

C. *ATRAX ROBUSTA* ENVENOMATION

The Sydney funnel-web spider is reputed to be the most dangerous Australian spider, but facts do not substantiate the claims.[33] Bites by the funnel-web spider may not be lethal to rabbits, guinea pigs, dogs, non-human primates, and cats.[34] There is little doubt that humans have died from the effects of the bite, which include muscle fasciculation, hypertension, tachycardia, excessive salivation and lacrimation, respiratory failure, and coma.[10] The signs noted are associated with catecholamine stimulation and may be caused by stimulation of endogenous catecholamine release by the venom.[33]

The pharmacologic response of an anesthetized dog and cat to intravenous funnel-web spider venom is that of transient hypertension, tachycardia, body temperature decrease, and increased cerebrospinal fluid pressure.[33]

D. *IXEUTICUS ROBUSTUS* ENVENOMATION

In one report of large animal spider envenomation, Pascoe[11] describes hot, edematous, and painful swellings about the head and neck of horses stabled in stalls known to be frequented by black house spiders.

E. OTHER SPIDER BITES

There are no documented cases of spider bites in animals from the dozens of other species that bite humans.

V. DIAGNOSIS

Diagnosing a spider bite without direct evidence of spider exposure is exceedingly difficult. Clinical signs may be extremely variable. Lymphocyte blastogenesis studies may be carried out in *Latrodectus* and *Loxosceles* species envenomation if specific antisera can be obtained. It may be necessary to collect acute and convalescent samples for comparison.[5]

A. NECROPSY

Superficial lesions and those found at necropsy are nonspecific. A necrotic dermal ulceration that is slow to heal is suggestive of brown recluse spider bite.

VI. PROGNOSIS

Spider bites are rarely fatal in humans and no fatal cases in animals have been documented.

VII. THERAPY

As most diagnoses of spider bite are subjective, a veterinarian usually treats the suspected bite site as an inflammatory weal, applying cold compresses and prescribing antihistamines and corticosteroid therapy. Systemic signs should be monitored and treated appropriately with fluids and other supportive medications.

A. *LATRODECTUS* BITES

An antivenin (Lyovac®, equine origin, Merck Sharp and Dohme) is available in the U.S. One vial, administered intravenously, is usually sufficient for a dog or cat.[19] If the history is suggestive of a black widow spider bite in a dog or cat, antivenin should be administered intravenously as quickly as possible. It is not necessary to conduct a horse serum skin sensitivity test in animals, but the patient should nonetheless be monitored for development of signs of anaphylaxis. Pretreatment with an antihistamine [10 mg/kg, diphenhydramine HCl (Benadryl®)] may be desirable to prevent horse serum anaphylaxis. Also be prepared to administer epinephrine intravenously at a dosage of 0.01 mg/kg [3 to 5 ml of 1:1000 (1 mg/ml) solution per 450-kg horse or cow]. The dosage for a small animal is 0.01 to 0.02 mg/kg [use a 1:10,000 dilution (0.1 mg/ml)]; for example, 1 ml/10-kg dog.

Acute muscular pain may be ameliorated with an intravenous infusion of calcium gluconate (1 ml/kg of a 10% solution). Calcium therapy may be repeated as necessary. Pain may also be controlled with meperidine HCl (Demerol®), 10 mg/kg dogs, 3 mg/kg cats, or diazepam (Valium®) 0.5 to 1.0 mg/kg.

B. *LOXOSCELES* BITE

In human medicine, brown recluse spider bite therapy is highly controversial.[5] Some physicians administer oral and/or intralesion corticosteroids, while others feel this is not indicated. Early surgical excision is advocated by some.[5] Berger[32] contends that recluse spider bites may be overtreated because some bites resolve themselves without progressing to the disfiguring necrotic dermal ulcerative stage. Corticosteroids are frequently administered to a patient with a suspected envenomation; however, steroids do not inactivate the venom nor do they block its primary action or lessen the cutaneous necrosis in *Loxosceles* bites.[5]

Although early surgical excision of the bite area seems logical, dispersion of the venom is rapid and may not be coincident with the area of necrosis, making surgical margins difficult to establish.[5,35] Numerous other medical treatments have been advocated at one time or another. In humans and in animals, systemic manifestations (shock, hemolysis, coagulopathies and dysuria) of recluse spider bite require immediate attention.

No antivenins are available in the U.S., although they have been used in South America.[5] Antivenin is unlikely to be of value in animals because antivenin must be administered within 30 min of the time of the bite to be efficacious; animal bites are rarely diagnosed until late in the course of development of the poisoning.

In animals, once the lesion reaches the necrotic ulceration stage, debridement and general wound care is indicated. Healing is prolonged. Skin grafting should not be attempted before 2 months have elapsed from the original bite, as an earlier graft will necrose and be rejected. Antibiotic therapy is advised with extensive or deep ulcerations.[19]

C. *ATRAX ROBUSTA* BITE

The venom of the funnel-web spider is nonproteinaceous and antivenins are not available. Supportive and local wound care are indicated.[11]

REFERENCES

1. **Harwood, R. F. and James, M. T.,** *Entomology in Human and Animal Health,* 7th ed., Macmillan, New York, 1979.
2. **Bucherl, W.,** Spiders, in *Venomous Animals and Their Venoms,* Vol. 3, Bucherl, W. and Buckley, E. E., Eds., Academic Press, Orlando, FL, 1971, 197.
3. **Ori, M.,** Biology of and poisonings by spiders, in *Handbook of Natural Toxins, Vol. 2. Insect Poisons, Allergens and Other Invertebrate Toxins,* Tu, A. T., Ed., Marcel Dekker, New York, 1984, 397.
4. **Kraus, O.,** The Arachnids and their relatives, in *Grzimek's Animal Life Encyclopedia,* Vol. 1, Grzimek, B., Ed., Van Nostrand Reinhold, New York, 1974. 403.
5. **Wong, R. C., Hughes, S. E., and Voorhees, J. J.,** Spider bites *Arch. Dermatol.,* 123, 98, 1987.
6. **Caras, R.,** Arachnids, in *Venomous Animals of the World,* Caras, R., Ed., Prentice-Hall, Englewood Cliffs, NJ, 1974, 48.
7. **Parrish, H. M.,** Analysis of 460 fatalities from venomous animals in the United States, *Am. J. Med. Sci.,* 245, 129, 1963.
8. **Rauber, A.,** Black widow spider bites, *J. Toxicol. Clin. Toxicol.,* 21, 473, 1983-1984.
9. **Maretic, Z. and Stanic, M.,** The health problem of arachnidism, *Bull. World Health Org.,* 11, 1007, 1954.
10. **Seawright, A.,** Spider bites, in *Diseases of Livestock,* 8th ed., Hungerford, T. G., Ed., McGraw-Hill, Sydney, 1975, 1087.
11. **Pascoe, R. R.,** The nature and treatment of skin conditions observed in horses in Queensland, *Austr. Vet. J.,* 49, 35, 1973.
12. **Ebeling, W.,** *Urban Entomology,* University of California, Division of Agricultural Sciences, Los Angeles, 1975.
13. **Geren, C. R. and Odell, G. V.,** The biochemistry of spider venoms, in *Handbook of Natural Toxins, Vol. 2. Insect Poisons, Allergens and Other Invertebrate Toxins,* Tu, A. T., Ed., Marcel Dekker, New York, 1984, 441.
14. **McCrone, J. D. and Hatala, R. J.,** Isolation and characterization of a lethal component from the venom of *Latrodectus mactans* in animal toxins, in *Animal Toxins,* Russell, F. E. and Saunders, P. R., Eds., Pergamon Press, Oxford, 1967.
15. **Duchen, L. W. and Gomez, S.,** Pharmacology of spider venoms, in *Handbook of Natural Toxins, Vol. 2. Insect Poisons, Allergens and Other Invertebrate Toxins,* Tu, A. T., Ed., Marcel Dekker, New York, 1984, 483.
16. **King, L. E., Jr.,** Spider bites, *Arch. Dermatol.,* 123, 41, 1987.
17. **Oehme, F. W., Brown, J. F., and Fowler, M. E.,** Toxins of animal origin, in *Casarett and Doull's Toxicology—The Basic Science of Poisons,* 2nd ed., Doull, J., Klassen, C. D., and Amdur, M. O., Eds., Macmillan, New York, 1980, 557.
18. **Brown, R. A.,** Red-back spider envenomation in dogs. Control and therapy, *No. 989, Post. Graduate Committee in Vet. Sci., Univ. of Sydney,* Sydney, Australia, 1980.
19. **Kirk, R. W., Bistner, S. I., and Ford, R. B.,** *Handbook of Veterinary Procedures and Emergency Treatment,* 5th ed., W. B. Saunders, Philadelphia, 1990, 198.

20. **Meerdink, G. L.,** Bites and stings of venomous animals, in *Current Veterinary Therapy, VIII. Small Animal Practice*, Kirk, R. W., Ed., W. B. Saunders, Philadelphia, 1983, 155.
21. **Mount, M. E.,** Biotoxins, in *Textbook of Veterinary Internal Medicine*, 3rd ed., Ettinger, S. J., Ed., W. B. Saunders, Philadelphia, 1989, 472.
22. **Clarke, M. L., Harvey, D. G., and Humphreys, D. J.,** *Veterinary Toxicology*, 2nd ed., 1981, 290.
23. **Kobert, R.,** *Beitrage zur Kenntinis der Giftspinnen*, Ferd. Enke. Verl., Stuttgart, 1901.
24. **Schtschensnowitsch, M.,** *Latrodectus* poisoning in animals including the camel, *St. Petersburger Med. Wchschr.*, 1870, 54.
25. **Schtscherbina, A.,** Serum als Heilmittel bei den Bissen der Karakurte/*Latrodectus maimignattus* (Serum as a treatment for bites by the black widow spider), *Arb. de Entomologie,* Bureau St. Petersburg, 2(No. 4), 1903.
26. **Maretic, Z.,** Latrodectism in Mediterranean countries, including South Russia, Israel and North Africa, in *Venomous Animals and Their Venoms,* Vol. 3, Bucherl, W. and Buckley, E. E., Eds., Academic Press, Orlando, FL, 1971, 299.
27. **Maretic, Z. and Haberhehl, G.,** Latrodektismus bei Menschen und Tieren sowie grosse Tierseuchen des vorigen Jahrhunderts hervorgerrufen durch Latrodektus (Lactrodectism in man and animals), *Dtsch. Tierartztl. Wschr.*, 92, 245, 1985.
28. **Roza, M. and Soedibio, R.,** Cattle envenomation by black widow spiders in India, *Hemera Zoa*, 18, 169, 1950.
29. **Vellard, J.,** *Le Venin des Araignrées*, Paris, 1936.
30. **Wasserman, G. S. and Siegel, C.,** Loxoscelism (brown recluse spider bites): a review of literature, *Vet. Human Toxicol.*, 19, 256, 1977.
31. **Muller, G. H., Kirk, R. W., and Scott, D. W.,** Spiders, in Small Animal Dermatology, 4th ed., Muller, G. H., Kirk, R. W., and Scott, D. W., Eds., W. B. Saunders, Philadelphia 1989, 407.
32. **Berger, R. S.,** The unremarkable brown recluse spider bite, *JAMA*, 225, 1109, 1973.
33. **Garnet, J. R.,** *Venomous Australian Animals Dangerous to Man*, Commonwealth Serum Laboratories, Parkville, Victoria, 1968.
34. **Tibbals, J., Sutherland, S. K., and Duncan, A. W.,** Effects of male Sydney funnel-web spider venom in a dog and a cat, *Austr. Vet. J.*, 64, 1987.
35. **Rees, R. S., Altenbern, D. P., Lynch, J. B., and King, L. E., Jr.,** Brown recluse spider bites: a comparison of early surgical excision versus Dapsone and delayed surgical excision, *Ann. Surg.*, 202, 658, 1985.

Chapter 8

SCORPIONS

I. IDENTIFICATION AND BIOLOGY

There are 650 species of scorpions (Table 1), that inhabit mild temperate to tropical climates throughout the world. They range in size from 2 to 20 cm.[1,2] Scorpions are predators, feeding on other scorpions, spiders, various insects, and worms.[1] Usually, prey is grasped with pincers and torn apart with the chelicerae. If the victim struggles excessively or is larger than the usual prey, a scorpion will sting it. However, the sting is a last resort, as a scorpion discharges all available venom in each sting and a number of days may be required to replenish the supply.

All scorpions produce venom, but only a few species are dangerous to people. Perhaps even fewer species cause clinical poisoning in animals.

A. MORPHOLOGY

Scorpions have four pairs of legs, as do spiders (see Figures 1 and 2). The pedipalps are modified and support pincers at the tip for grasping prey. The terminal abdominal segment (telson) contains two venom glands with small ducts leading to two separate orifices on the stinger (Figure 2, inset).[1] Scorpions have eyes, but are virtually blind, depending on tactile senses that operate through special organs (pectines) located on the lateral abdomen just caudal to the last pair of legs.[1]

B. DISTRIBUTION

Though scorpions require water to drink, they may inhabit arid deserts by capitalizing on nocturnal behavior and the ability to estivate for long periods of time (up to 6 months).[3] Dangerous species are primarily arid climate dwellers.

The bark scorpion, *Centruroides sculpturatus,* is the most dangerous species in the U.S. and is unique in that it is a climber, preferring the bark of Arizona desert trees to burrows for hiding places.[4] This species in found primarily in Arizona, but may also be found in parts of Texas, New Mexico, northern Mexico, and a few small areas in California.[4] It is appropriate to mention that animals and humans may be envenomated by such species outside native distribution areas because tourists may inadvertently carry scorpions to distant places in their belongings.[4]

II. CONDITIONS OF POISONING

Scorpions are nocturnal and prefer to live where moisture is available. The more venomous species are found in the southwestern U.S. Arizona reports the highest prevalence of scorpion stings in humans in the U.S.[5] Animal stings are likely to be highest in these areas as well, but no statistics are available.

TABLE 1
Classification of the Scorpions

Phylum — *Arthropoda*
Class — *Arachnida*
Order — *Scorpionida*
  Family — *Buthidae*
    Centruroides sculpturatus — Arizona
    Centruroides gertschi — Arizona, southern California, Mexico
    Centruroides spp. — Central America and West Indies
    Tityus spp. — Mexico, South America, West Indies
    Hardrurus spp. — Mexico and southern states of U.S.
    In the Old World, a number of *Buthidae* genera have been implicated in human envenomation including:
    Heterometrus spp. — India
    Androctonus spp. — India to eastern and southern Mediterranean region west to Morocco
    Buthacus spp. — Atlantic coasts to Palestine, Senegal, and Tunis
    Leiurus spp. — Syria, Palestine, Egypt, Yemen
    Buthotus spp. — Palestine, Syria, Algeria, Morocco
    Buthus spp. — North Africa, Somalia, Ethiopia, parts of Palestine, Spain, France
    Parabuthus spp. — South Africa to Sudan
    Hardrurus spp. — Mexico and southern states of U.S.
  Family — *Scorpionidae*
    Pandinus spp. — Africa and Arabia
    Scorpio spp. — North Africa
  Family — *Vejovidae*
    Vejovis spp. — from British Columbia, Canada to Mexico

From Bucherl, W., *Venomous Animals and Their Venoms*, Bucherl, W. and Buckley, E. E., Eds., Academic Press, Orlando, FL, 1971, 317; Tu, A. T., *Venoms: Chemistry and Molecular Biology*, Tu, A. T., Ed., John Wiley & Sons, New York, 1977, 459. With permission.

Scorpions may be grouped into species that inhabit fields or live remote from human habitation and those that readily move into houses and buildings, finding refuge in attics, closets, basements, garages, under floor crawl spaces, stacks of old lumber, porches, sacks, couches, beds, piles of stored newspapers, or even in shoes and clothing. Field scorpions live in burrows, some of which are elaborate, with chambers designed to provide varying degrees of temperature.[6] Scorpions may also establish burrows in children's sandboxes, under rocks, and in association with ant and termite mounds.[6]

Pet exposure to scorpions is likely to be accidental, similar to that of humans, particularly children. The movement of a sleeping pet may initiate a stinging response. There have been no reports of scorpion stings in livestock or horses.

III. VENOM

Scorpion venom is complex, consisting of a mixture of many pharmacologically active, neurotoxic proteins (as many as 16 different proteins in a single venom).[7,8] Some proteins are enzymes, others not. Phospholipase A_2 is com-

FIGURE 1. Scorpions from Wyoming. Identity unknown.

monly present in the venom of scorpions, while other enzymes such as hyaluronidase, phosphomonoesterase, acetylcholinesterase, and others may be found in the venom of some species but not in others.[7]

Although proteins predominate in scorpion venoms, non-protein constituents are found in variable concentrations, including amino acids, histamine, 5-hydroxytryptamine, and serotonin.[7] The contribution of each of these non-protein constituents to the toxicity of a venom is not precisely known.

Scorpion venoms are antigenic, with some closely related species cross-reacting with each other. Antivenins are produced for the more highly venomous species.[7]

Scorpion venom causes release of acetylcholine at the motor end-plate of muscle, preventing normal nerve impulse transmission and causing muscle spasms. Cardiac-related signs of scorpion envenomation, such as hypertension, tachycardia, and pulmonary edema, are probably results of the release of catecholamines from the adrenergic nerve endings.[7,9,10]

IV. CLINICAL SIGNS

Scorpions may be classified into two groups depending on the nature of the effect of the venom on human beings.[11] The first group (the vast majority of scorpions) cause only a mild to severe local reaction at the sting site. The initial sting is associated with a sharp burning sensation followed by a slight swelling, in some cases progressing to a severe, painful swelling with ecchymosis. Tissue necrosis does not occur.[11]

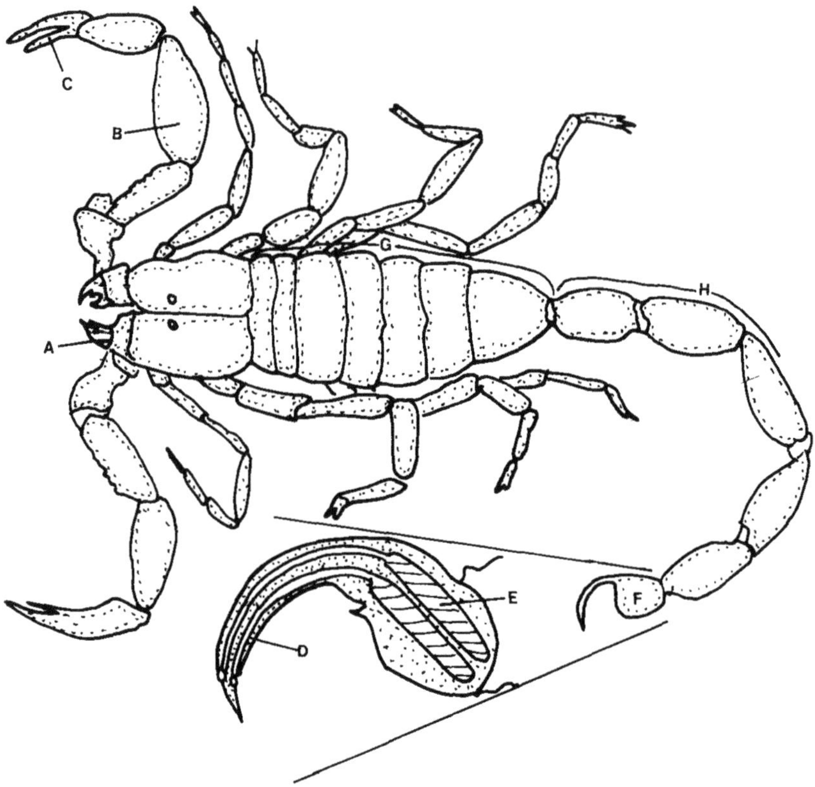

FIGURE 2. Diagram of a scorpion; inset, telson and stinger. A, chelicera; B, pedipalp; C, pincer; D, stinger; E, venom gland; F, telson; G, preabdominal segments; H, post-abdominal segments.

The second group of scorpions produces venom that may be lethal to human beings. Species in this group found in the U.S. include *Centruroides sculpturatus, C. gertschi, Vejovis spinigerus,* and *Hardrurus arizonensis.* A sting from one of these scorpions produces no visible reaction at the point of injection, but there is usually local pain (tapping on the bite site — called the tap test — produces a sharp pain),[4] and the victim may experience hyperesthesia. The venom is primarily neurotoxic and variable symptoms have been reported, including numbness around the wound which may spread to involve an entire limb, difficulty in swallowing, "thick-tongue" feeling, sweating, headaches, dizziness, dyspnea, excessive salivation, gastric tympany, alternating episthotonos and opisthotonus spasms, mydriasis, nystagmus, diplopia, temporary blindness, incontinence, involuntary defecation, penile erection, hypertension, respiratory arrest, and cardiac failure.[7,11]

Curry et al.[4] state, "Cranial nerve dysfunction is characterized by roving and/or rotatory eye movements, blurred vision, fasciculation of the tongue and loss of control of the pharyngeal muscles with resultant stridor and occasionally respiratory arrest. Excessive somatic neuromuscular activity presents initially as uncontrollable jerking of the extremities, or what appears to be restlessness, with the victim unable to lie still. This then may progress to chaotic thrashing about."

Victims that live for a few hours may develop hyperkalemia and hyponatremia. Death from cardiac failure and pulmonary edema may occur in less than 1 h or in several days.[7]

Pets may experience stings from mildly venomous species without the owner ever noticing the local reaction which usually subsides within a few hours.[12,13] It would be difficult to differentiate a scorpion sting from that of a wasp or yellow jacket in the early stages.

Systemic signs are similar to those of a parasympathomimetic reaction and include salivation, muscle fasciculation, generalized weakness, and paralysis including respiratory paralysis. The clinical condition usually improves within 24 h.[12,13]

The pharmacophysiologic effects of *Centruroides sculpturatus* venom administered intravenously to anesthetized dogs and cats was hypertension, respiratory failure, and skeletal muscle stimulation. Anoxia was the terminal result.[14]

V. DIAGNOSIS

Definitive diagnosis of scorpion sting in an animal is rare. Veterinarians practicing in regions inhabited by highly venomous scorpions treat as stung animals that exhibit several of the signs recognized for scorpion sting in people.

Myocardial necrosis is a consistent lesion found in fatal human stings. Pulmonary edema is a common finding and may be associated with cardiomyopathy or develop terminally as a result of anoxic struggling.

VI. TREATMENT

The severity of the local reaction may be inhibited in the early stages by application of cold compresses to the lesion. Freezing (ligation and cryotherapy)[11] is no longer an acceptable form of therapy for venomous bites and stings in human beings,[15] and should not be employed in animal bites either. Other unacceptable methods of therapy include incision and suction, or application of a tourniquet.

Antivenin is available for *Centruroides stings*,[3] but its use in animal stings is questionable because of the delay in beginning treatment in most cases. Questions about the availability and use of scorpion antivenins should be referred to the nearest Poison Control Center.

Most stings, even from highly venomous species, may be treated by atropine, 0.04 mg/kg (or to effect), to counter the parasympathomimetic effects. Corticosteroids are frequently administered to decrease shock and edema. The administration of parenteral fluids must be carefully monitored to avoid the risk of contributing to pulmonary edema. Paralysis of respiratory muscles may necessitate positive pressure respiratory assistance.[12,13,16]

Meperidine HCl (Demerol®), and probably other narcotics, are contra-indicated in the treatment of scorpion stings because clinical and experimental evidence suggests that narcotics act synergistically to increase toxicity of the venom.[7]

VII. PROGNOSIS

Fatality of animals is rare with scorpion stings. Signs usually improve within 24 h.

VIII. PREVENTION

Prevention is difficult because of the ubiquitous sites in which scorpions conceal themselves in the daytime and their wanderings at night. It would be desirable to eliminate hiding places, but this is rarely possible in a household. Severe scorpion infestations may be controlled by insecticide applications. Insecticides should be applied by professional pest control operators or under the direction of local Agricultural Commissioners or County Cooperative Extension specialists.

REFERENCES

1. **Caras, R.,** *Venomous Animals of the World*, Prentice-Hall, Englewood Cliffs, NJ, 1974.
2. **Harwood, R. F. and James, M. T.,** *Entomology in Human and Animal Health*, 7th ed., Macmillan, New York, 1979, 452.
3. **Ebeling, W.,** Scorpions, in *Urban Entomology*, University of California, Division Agricultural Science, Los Angeles, 1975, 343.
4. **Curry, S. C., Vance, M. V., Ryan, P. J., Kunkel, D. B., and Northey, W. T.,** Envenomation by the scorpion *Centruroides sculpturatus*, *J. Toxicol. Clin. Toxicol.*, 21, 417, 1983–1984.
5. **Parrish, H. M.,** Analysis of 460 fatalities from venomous animals in the United States, *Am. J. Med. Sci.*, 245, 129, 1963.
6. **Bucherl, W.,** Classification, biology and venom extraction of scorpions, in *Venomous Animals and Their Venoms*, Bucherl, W. and Buckley, E. E., Eds., Academic Press, Orlando, FL, 1971, 317.
7. **Tu, A. T.,** Scorpion venom, in *Venoms: Chemistry and Molecular Biology*, Tu, A. T., Ed., John Wiley & Sons, New York, 1977, 459.
8. **Possani, L. D.,** Structure of scorpion toxins, in *Handbook of Natural Toxins, Vol. 2. Insect Poisons, Allergens, and Other Invertebrate Venoms*, Tu, A. T., Ed., Marcel Dekker, New York, 1984, 513.
9. **El-Asmar, M. F.,** Metabolic effect of scorpion venom, in *Handbook of Natural Toxins, Vol. 2. Insect Poisons, Allergens, and Other Invertebrate Venoms*, Tu, A. T., Ed., Marcel Dekker, New York, 1984, 551.

10. **Couraud, F. and Jover E.,** Mechanism of action of scorpion toxins, in *Handbook of Natural Toxins, Vol. 2. Insect Poisons, Allergens, and Other Invertebrate Venoms,* Tu, A. T., Ed., Marcel Dekker, New York, 1984, 659.
11. **Stahnke, H. L.,** *The Treatment of Venomous Bites and Stings,* Revised ed., Arizona State University, Tempe, 1966.
12. **Meerdink, G. L.,** Bites and stings of venomous animals, in *Current Veterinary Therapy, VII. Small Animal Practice,* Kirk, R. W., Ed., W. B. Saunders, Philadelphia, 1980, 174.
13. **Meerdink, G. L.,** Bites and stings of venomous animals, in *Current Veterinary Therapy, VIII. Small Animal Practice,* Kirk, R. W., Ed., W. B. Saunders, Philadelphia, 1983, 155.
14. **Patterson, R. A.,** Physiological action of scorpion venom, *Am. J. Trop. Med. Hyg.,* 9, 410, 1960.
15. **Russell, F. E. and Saunders, P. R.,** *Animal Toxins,* Pergamon Press, New York, 1966.
16. **Oehme, F. W., Brown, J. F., and Fowler, M. E.,** Toxins of animal origin, in *Casarett and Doull's Toxicology — The Basic Science of Poisons,* 2nd ed., Doull, J., Klassen, C. D., and Amdur, M. O., Eds., Macmillan, New York, 1980, 557.

Chapter 9

TICK PARALYSIS

I. ETIOLOGY

Tick paralysis has been recognized for over 50 years; 43 species of ticks in 10 different genera have been incriminated.[1] Usually, only female ticks produce toxin which localizes in the saliva and is injected into the host when the tick takes a blood meal. However, nymphs and larvae of *Ixodes holocyclus* in Australia, *Argas persicus* in Africa, and male *Dermacentor andersoni, Hyalomma truncatum,* and *Rhipicephalus simus* in South Africa have been responsible for tick paralysis in human beings.[2] In the eastern U.S., *Dermacentor variabilis,* the common or American wood tick, is often involved. In the Rocky Mountain states and the Pacific Northwest, the Rocky Mountain tick *(Dermacentor andersoni)* and in California, the Pacific Coast tick *(Dermacentor occidentalis)* are the primary cause of paralysis in livestock. *Dermacentor variabilis* is more often the cause of paralysis in dogs.

II. CONDITIONS OF POISONING

A. HOSTS

Tick paralysis has been reported in many species of mammals and birds, but not in reptiles. Cases have been reported in horses, livestock,[3] llamas, dogs, and cats.[4] Wildlife species known to be affected include bison *(Bison bison),*[5] grey fox *(Urocyon cinereoargenteus),*[6] harvest mouse *(Reithrodontomys megalotis),*[7] ground hogs *(Marmota flaviventris avara),*[8] and black-tailed deer *(Odocoileus hemionus).*[4,9,10] Ticks causing paralysis are listed by geographical regions and reported hosts in Table 1.[4]

Host susceptibility varies. There may also be seasonal or annual variability as there have been reports of outbreaks in some years and none in others.

B. DISTRIBUTION

Tick paralysis has occurred in North America, Europe, Africa, Australia, and what was the U.S.S.R. Populations of ticks that produce the toxin seem to occur in pockets and are not spread uniformly throughout a region.

III. TOXIN

The precise chemical structure of the toxin is not known; however, in most species it is a neurotoxin that interferes with acetylcholine synthesis or liberation at neuromuscular endings, causing a lower motor neuron paresis and paralysis.[1] The bite of a single tick may cause paralysis in a human and a llama. A temporary immune response is associated with recovery from paralysis, but it is short-lived (2 to 5 weeks).

TABLE 1
Ticks Causing Paralysis in Specific Hosts, Listed by Geographical Region

Phylum — *Arthropoda*
Class — *Arachnida*
Order — *Acarina* (ticks and mites)

Region	Tick	Hosts affected
North America	*Dermacentor andersoni*	Humans, deer, llama, sheep
	D. variabilis	Dogs
	D. occidentalis	Cattle, horses
	Otobius megnini	Any
Europe	*Haemaphysalis inermis*	Humans, cattle, sheep
	H. punctata	Goats
	Hyalomma scupense	
Former U.S.S.R.	*Ixodes crenulatus*	Humans
	Ornithodoros lahorensis	
Africa	*Argus walkerae*	Chicken, cattle, humans
	Ixodes rubicundus	
	Rhipicephalus evertsi	
	Hyalomma truncatum	
	Argus arboreus	
West Africa	*Ornithodoros savigni*	Humans
Australia	*Ixodes holocyclus*	Humans, dogs, sheep, bandicoots

The mechanism of action of the toxin of the Australian tick, *Ixodes holocyclus,* differs from that of the toxin of *Dermacentor* spp. in that the action appears to be on central nerve centers rather than on peripheral nerves. Recovery following removal of the tick is prolonged.

IV. CLINICAL SIGNS

Tick paralysis in humans progresses as follows: signs are not noted until 5 to 7 d after the tick has started to feed. Initially, there is paresis of the legs, progressing to ataxia and total loss of motor function. Pain perception remains. Signs may progress rapidly, in a few hours, or more slowly for 24 to 48 h. Paralysis ascends to the torso, arms, neck, throat, and face, and the patient becomes tetraplegic. There may also be nystagmus, slurring of speech, dyspnea, and difficulty in chewing and swallowing. Body temperature is normal, as are hematologic and serum chemistry values. Death is caused by respiratory failure, when respiratory centers become affected.[11]

Signs of tick paralysis in animals are similar to those seen in humans. A grey fox was found with paresis and weak motor reflexes of the hind limbs, evidenced by an unsteady gait, knuckling, and ataxia.[6] In a terminal case, hind limb paresis was followed by an ascending flaccid paralysis involving the neck, throat, and face with associated difficulties in chewing, swallowing, and breathing. Ultimately, decreased sensibility and death ensued.

Similar signs have been reported in domestic animals. Many times, the only presenting sign may be total paralysis. In the early stages of tick paralysis, the animal is bright, alert, and able to eat and drink if the head and neck are supported appropriately.

V. DIAGNOSIS

Diagnosis is difficult unless a tick is found. Tick paralysis should be a factor in a differential diagnosis of any animal that is para- or tetraplegic, yet is bright and alert. No analytic method to identify the toxin is available. Response to removal of the tick(s) is usually dramatic.

A. NECROPSY

There may be no lesions at necropsy. Ticks may be found in the ears, crevices of the skin, and in one female human infant, the tick was found inside the vulva.[4]

VI. TREATMENT

Removal of all the ticks produces full recovery in a few minutes to days, if bulbar paralysis has not occurred. Total body application of insecticide sprays, powders, or dips may be effective if the insecticide reaches the tick. Ivermectin administered subcutaneously may kill imbedded ticks, but not rapidly enough to be useful in saving the life of an affected animal.

Heavily fibered animals such as llamas present a special challenge. Areas that are relatively fiberless, such as the axillary and inguinal areas around the face and ears and the perineum, should be searched before recommending clipping the body fiber.

Imbedded ticks must be removed carefully to avoid expressing additional toxin into the bite wound or leaving the head imbedded in the skin. Application of a recently extinguished match head to the caudal aspect of the tick may encourage it to release itself. Otherwise, fine tweezers or mosquito forceps may be used to grasp the tick as close to the skin as possible. Pull slowly at right angles to the skin to remove it.

VII. PROGNOSIS

If tick paralysis is diagnosed before bulbar paralysis occurs and the tick(s) are removed, recovery is uneventful in all locations except Australia. Paralysis caused by *Ixodes holocyclus* involves a long recovery period.

VIII. PREVENTION

A manager may attempt to keep highly susceptible species away from high-risk areas. It has been recommended that a preventive dose of ivermectin (0.4 mg/kg, subcutaneously) be administered before horses or llamas are taken on packing trips into tick-infested areas. As ivermectin has been reported to produce a residual effect for 3 weeks, this seems logical. Unfortunately, llama packers have reported that this method of prevention is not 100% effective.*

Packers or hikers in tick-infested areas should check themselves daily for ticks. It would be appropriate to do likewise for animals on the trek.

REFERENCES

1. **Gothe, R., Kunze, K., and Hoogstraal, H.,** The mechanisms of pathogenicity in the tick paralyses — A review, *J. Med. Entomol.*, 16, 357, 1979.
2. **Strickland, R. K. and Diamant, G.,** Ticks of veterinary importance, *U.S. Department of Agriculture Handbook No. 485*, 1965, 6.
3. **Loomis, E. C. and Bushnell, R. B.,** Tick paralysis in California livestock, *Am. J. Vet. Res.*, 29, 1089, 1968.
4. **Fowler, M. E.,** Tick paralysis, *Calif. Vet.*, 39, 25, 1985.
5. **Kohls, G. M. and Kramis, N. J.,** Tick paralysis in the American buffalo *Bison bison, Northwest Sci.*, 26, 61, 1952.
6. **Jessup, D. A.,** Tick paralysis in a grey fox *Urocyon cinereoargenteus, J. Wildlife Dis.*, 15, 271, 1979.
7. **Butzler, R. G., Albrecht, J., and Schaefer, T.,** Tick paralysis in a western harvest mouse *Reithrodontomys megalotis, J. Wildlife Dis.*, 16, 223, 1980.
8. **Gregson, J. D.,** Tick paralysis in ground hogs, Guinea pigs and hamsters, *Can. J. Comparative Med.*, 23, 266, 1959.
9. **Brunetti, O.,** Tick paralysis in California deer, *Calif. Fish Game Bull.*, 51, 208, 1965.
10. **Wilkinson, P. R.,** A first record of paralysis of a deer by *Dermacentor andersoni* and notes on the host potential of deer in British Columbia, *Proc. Entomolog. Soc.*, 62, 28, 1965.

* Markham, D., personal communication, 1991.

Chapter 10

MISCELLANEOUS ARTHROPODS

I. CENTIPEDES

Centipedes are segmented terrestrial arthropods. Numerous species are distributed throughout tropical and subtropical regions of the world, generally living on the forest floor, though a few species have adapted to human habitation. Centipedes have one pair of legs for each segment (15+). Species in the order *Scolopendromorpha* are all venomous and a few may inflict painful bites in humans and animals with the venom jaws (telopodites) associated with the cranial segment.[1]

Little is known about the venom of centipedes, but proteins in the form of enzymes have been isolated (endopeptidase) and also a cardiotoxic protein (toxin-S).[2] Non-protein constituents include the biogenic amines 5-hydroxytryptamine (serotonin) and histamine, plus lipids and polysaccharides.[2] Venoms with either coagulant or anticoagulant factors have been isolated from centipedes.[2]

In human beings, bites from larger species have been reported to produce a burning sensation, itching, erythema, mild swelling, and superficial necrosis at the bite site.[3] Systemic signs also have been reported in humans, including anxiety, vomiting, irregular pulse, dizziness, severe headache, and even death, but only rarely.[1] It may be difficult to separate fact from legend as far as centipede envenomation is concerned.

Companion animals may be envenomated by centipedes under the same circumstances in which humans come in contact with them. Envenomation is apparently mild because no case of animal bite by centipede has been documented in the literature.

II. MILLIPEDES

Millipedes do not bite and are not venomous, but they do have a glandular system that produces a foul-smelling, disagreeable fluid containing phenols and hydrocyanic acid. When the fluid comes in contact with the human eye, it causes a burning sensation and also may blister the skin, resulting in a wound that is slow to heal.[1] It is not known whether animals become affected in the same manner.

III. INSECTS

Venomous insects of major significance are described elsewhere in this book. Many other species, in either larval or adult stages, contain unpalatable substances that tend to dissuade predation, and which, in unique circumstances, may cause poisoning. Another group of insects (mosquitos, gnats, and midges) produces saliva that is irritating, causing a local reaction at the penetration site. In this sense, they are venomous.

A. MONARCH BUTTERFLY

Studies on the monarch butterfly, *Danaus plexippus*, during the 1970s were the first to demonstrate the ecologic significance of poisonous substances in the plant and animal kingdom as defenses against predation. Monarch butterflies have a typical lepidopteran life cycle (metamorphosis). The adult female lays her eggs on milkweeds, *Asclepias* spp. When the eggs hatch, larvae (caterpillars) feed on the leaves of the milkweed and develop through a number of instars before pupating into an inactive stage called the chrysalis (pupa). The adult butterfly emerges from the chrysalis and feeds on nectar and pollens.

When monarch caterpillars feed on milkweed leaves, they automatically consume the plant secondary compound (plant secondary metabolite) that is repugnant to insects other than the monarch and a milkweed beetle, *Chrysochus cobaltinus*. The monarch caterpillar is not affected by the toxin, but sequesters the toxin within its body, discouraging predation by insectivores (primarily birds). The toxin is transferred to the chrysalis and the adult butterfly.

The monarch caterpillar is brightly colored (green with black and yellow stripes) and easily recognized. The adult monarch is orange and black and it, too, is readily recognized and avoided by would-be predators. The cardioactive glycoside toxin causes immediate vomition in the naive bird, which is a positive stimulus for that bird to avoid butterfly or caterpillar in the future. The story could end here, serving as an excellent example of a commensal relationship between the butterfly, its larvae, and the milkweed plant. However, with far-reaching effects, other lepidopterans such as the viceroy butterfly, *Limenitis archippus*, have employed the principle of mimicry to capitalize on the monarch's protective coloration, thus also avoiding predation.

Since the story of the monarch butterfly has unfolded, numerous other examples of toxic plant/insect interactions have been identified. Insectivorous wild mammals and birds learn to avoid unpalatable or toxic species. It is likely that fatalities in wild mammals or birds are rare, but could occur.

B. SAWFLY LARVAE

The poisonous effects following ingestion of an Australian sawfly, *Lophyrotoma interrupta*, have been known since 1932[4] and continue to cause death in cattle and sheep.[5,6] The insect is found throughout eastern Australia, but poisoning has been restricted to grazing areas associated with forests of silver-leaf ironbark eucalyptus, *Eucalyptus melanophloia*, that support periodic plague numbers of the insects.[7] The larvae accumulate in large piles under the host trees and are eaten by cattle and sheep. The larvae contain an octapeptide (lophyratomin) that causes hepatic and renal degeneration. The clinical syndrome is characteristic of renal or hepatic insufficiency. The severity and course of poisoning depends on the ingested dose. Depression, polyuria, mania, and coma have been observed.[5,8]

C. BUTTERFLIES

Over 160,000 named species of moths and butterflies in the insect order *Lepidoptera* (scaly wings) range throughout tropical, subtropical, and temperate regions of the world. The larvae are commonly called caterpillars (derived from the Latin "*catta pilosa*" or hairy cat).[9] The vast majority of species of butterflies and moths are inoffensive, but approximately 200 species are known to cause human allergic respiratory reactions to hairs or scales, and dermatitis as a result of contact with hairy caterpillars (Figure 1).[9] The species that are likely to cause health problems in humans are found in the following genera: *Euproctis, Thaumetapoea, Hylesia, Automeris, Megalopyge,* and *Latoia.*

Disease caused by butterflies is called *lepidopterism.* Some researchers restrict this term to only those cases involving adult butterflies, and use the term "erucism" (Latin: *eruca* = caterpillar) for cases involving the larvae.[9]

Several types of hair cover the bodies of butterflies, moths, and other insects. The hairs serve various functions, one of which is to deter predation. Insect hair is produced by modification of specific epidermal cells. Hard, stout hairs are called setae or bristles; more flattened projections are called scales.[9] Some hairs (spicule hairs) and scales are easily detached from the insect and may become aerosolized and inhaled by animals. Venomous spicule hairs are called "flechettes" by some authors.[10]

Venomous species also have spine hairs associated with glands that produce venom. Spine hairs must remain attached to the insect's nervous system in order for venom to be extruded after the tip of the spine has been broken off.[9]

The following clinical syndromes have been described in the literature on human envenomation.

Dermatologic syndrome: Penetration of the skin by spicule hairs and nonvenomous setae may produce mild irritation and itching, typical of reaction to a foreign body, and associated with histamine release. A more severe form of dermal reaction is caused by exposure to poison-bearing spines. Injected venom causes an immediate sharp pain, as from a needle puncture, and the subsequent reaction is more intense.

Within a few minutes to several hours following the appearance of initial signs, an area of erythema and whitish edematous papules develop that may progress to an urticarial weal surrounded by smaller plaques. Generally, the reaction subsides in 12 to 14 h, but it may persist for several days. The vesicles may rupture, resulting in open wounds and scab formation.[9]

Ophthalmic injuries: Spicule hairs detach easily and may float in the air. If the spicules enter the conjunctival sac, they may produce conjunctivitis and keratitis, causing photophobia, increased lacrimation, erythema, and blepheredema.

Respiratory and digestive syndromes: Spicule hairs may be inhaled, causing rhinopharyngitis, bronchitis, coughing, sneezing, rhinorrhea, and dyspnea. Ingestion of the spicules causes stomatitis, enteritis, and even death.

Lepidopterism occurs in livestock and pets. In South Africa, deaths of pigs and cattle have been reported as a result of the ingestion of caterpillars of *Nudaurelia* sp. Severe stomatitis has been reported when dogs have consumed

FIGURE 1. Urticarial caterpillars from Brazil, species unknown.

grass contaminated with the shed spicule hairs of caterpillars of *Thaumetopoea pinivora*. In Colorado and New Mexico, the range caterpillar, *Hemileuca oliviae*, not only consumes valuable forage, but may also cause stomatitis in cattle that inadvertently ingest them. European ducks developed severe enteritis after ingesting unidentified caterpillars.[11] The milder effects of urticarial butterflies and moths are likely to go undetected.

Treatment of lepidopterism in animals is usually local application of cold packs for relief of dermatologic conditions. The administration of mineral oil may be soothing if larvae have been ingested. Antihistamines have provided relief in human victims.

D. FLIES, MOSQUITOS, AND GNATS

The order *Diptera* includes thousands of species that are annoying or parasitic to humans, and domestic and wild animals. Dipterans are also important disseminators of infectious microorganisms. It is sometimes difficult to differentiate venomous from parasitic species. The saliva of hematogenous dipterans may contain a toxin that aids in altering the endothelium of capillaries to induce free flow of blood which is then ingested by the fly.

Many humans are extremely sensitive to the bites of such flies and develop erythema, urticarial weals, and, with scratching that follows the bite, the epithelium may be stripped off, leaving an open wound.

Animals may likewise be sensitive to the toxin in the saliva. Even non-biting species, such as the house fly *(Musca domestica)* and face fly *(Musca autumnalis)*, feeding on lacrimal secretions or open wounds may produce a local reaction causing blepharitis, conjunctivitis, and keratitis.[12] Biting flies, mosquitos, gnats, and midges (see Table 1) may produce local reactions in animals as well as humans. An attack by immense numbers of simulid gnats may cause debility and even death in livestock.[11] Horses, mules, and cattle were killed by massive attacks of the buffalo gnat *(Cnephia pecuarum)* in the Mississippi valley.[11]

Domestic and wild animals living in areas heavily infested with biting dipterans often employ behavioral traits to minimize exposure (crowding together, head shaking, tail switching, smearing themselves with mud, and running). As a youngster, the author was mowing alfalfa hay using a team of horses. A cloud of simulid gnats descended upon them. It was difficult to control the horses from bolting and at the same time protect oneself from the biting insects.

TABLE 1
Classification of Miscellaneous Venomous Arthropods

- Phylum — *Arthropoda* — jointed-legged invertebrates
- Class — *Chilopoda*
- Order — *Scolopentromorpha*
- Family — *Scolopenridae*
  - *Scolopendra subspiripes* — large tropical centipedes
  - *Scolopendra viridicornis* — large tropical centipedes
- Class — *Diplopoda* — millipedes
  - *Rhinocricus lethifer* — Haitian millipede
- Class — *Insecta* — insects
- Order — *Lepidoptera* — butterflies, moths
  - *Cerura* sp. — puss moth caterpillar
  - *Zygaena* spp. —
  - *Nudaurelia cytherea* — South Africa
  - *Thaumetopoea pinivora* —
  - *Hemileuca oliviae* — range caterpillar of Colorado and New Mexico
- Order — *Diptera* — flies, mosquitos, gnats
- Family — *Simuliidae* — black flies, gnats
  - *Simulium* spp. —
  - *Cnephia pecuarum* — buffalo gnat
- Family — *Ceratopogonidae* — biting midges
  - *Culicoides* spp. — "punkies," "no-see-ums," and "sand flies"
- Family — *Culicidae* — mosquitos
  - Many genera including *Aedes, Anopheles*, and *Culex*
- Family — *Muscidae* — flies
  - *Musca domestica* — house fly, does not bite but annoys
  - *Musca autumnalis* — face fly, does not bite
  - *Stomoxys calcitrans* — biting stable fly
  - *Lophyrotoma interrupta* — sawfly (larvae)
- Family — *Tabanidae* — horse flies, deer flies
  - *Tabanus* spp. — horse flies
  - *Chrysops* spp. — deer flies
  - *Euproctic flava* —

REFERENCES

1. **Caras, R.**, *Venomous Animals of the World*, Prentice-Hall, Englewood Cliffs, NJ, 1974, 73 and 92.
2. **Jangi, B. S.**, Centipede venoms and poisoning, in *Handbook of Natural Toxins, Vol. 2. Insect Poisons, Allergens, and Other Invertebrate Venoms*, Tu, A. T., Ed., Marcel Dekker, New York, 1984, 333.
3. **Bucherl, W.**, Venomous chilopods or centipedes, in *Venomous Animals and Their Venoms, Vol. 3. Venomous Invertebrates*, Bucherl, W. and Buckley, E. E. Eds., Academic Press, Orlando, FL, 1971, 169.
4. **Roberts, F. H. S.**, The cattle-poisoning sawfly *Pterygophoeus analis, Queensland Agric. J.*, 37, 41, 1932.
5. **Calloway, L. L.**, Sawfly poisoning in cattle, *Queensland Agric. J.*, 81, 155, 1955.

6. **Dadswell, L. P., et al.**, Sawfly larvae, *Austr. Vet. J.*, 62, 94, 1985.
7. **Leonard, G. J.**, The isolation of a toxic factor from sawfly *Lophyrotoma interrupta* larvae, *Toxicon*, 10, 597, 1972.
8. **Blood, D. C. and Radostits, O. M.**, *Veterinary Medicine*, 7th ed., Bailliere Tindall, London, 1989, 1358.
9. **Kawamoto, F. and Kumada, N.**, Biology and venoms of *Lepidoptera*, in *Handbook of Natural Toxins, Vol. 2. Insect Poisons, Allergens, and Other Invertebrate Venoms*, Tu, A. T., Ed., Marcel Dekker, New York, 1984, 291.
10. **Pesce, H. and Delgado, A.**, Poisoning from adult moths and caterpillars, in *Venomous Animals and Their Venoms, Vol. 3. Venomous Invertebrates*, Bucherl, W. and Buckley, E. E., Eds., Academic Press, Orlando, FL, 1971, 119.
11. **Harwood, R. F. and James, M. T.**, Urticating lepidoptera, in *Entomology in Human and Animal Health*, 7th ed, 1979, 436.
12. **Pascoe, R. R.**, *A Colour Atlas of Equine Dermatology*, Wolfe Publishing, Ipswich, England, 1989, 250.

PART IV: Vertebrates

Chapter 11

AMPHIBIAN POISONING

I. ETIOLOGY

A. IDENTIFICATION

The poisonous amphibians discussed in this chapter are listed in Table 1. Amphibians were the first vertebrates to bridge the gap between aquatic and terrestrial environments and a major challenge was to cope with desiccation.[1-5] In both aquatic and terrestrial environments, the skin functions as a respiratory organ as well as a protective membrane. To prevent desiccation and maintain respiratory exchange, the skin is bathed with secretions from skin glands.[1] Some of these secretions also serve to discourage predation.

There are two types of glands: *Mucous glands* are broadly distributed over the entire body.[1-6] The secretion is fluid and slimy, keeping the skin moist and lubricated, particularly in aquatic and semiaquatic species.[5] When these species are grasped for restraint, struggling stimulates production of additional secretion to facilitate escape. Mucous gland secretion may be odoriferous and/or bad tasting to deter predation.

Granular glands are more restricted in distribution on the body (sides of the head, shoulders, and dorsolateral edges). The secretion is generally thicker and creamier than that of the mucous glands.[5,6] Toads tend to have more granular glands than frogs, and in the poisonous species, an aggregation of these glands is found caudal and lateral to the ear, coalesced into what is known as the *parotid gland.*[6]

Confusion arises as to the proper terminology for this gland because of inference to the parotid salivary gland of mammals. It is spelled in a variety of ways: parotid, parotoid, and paraotoid. The roots of the words are: "para" (par) = beside or near; "otid" = ear; "oid" = like or similar to. As the gland in amphibians is a skin gland and has no similarity to the salivary gland of a mammal, the spelling "parotoid" seems improper. The term "parotid" describes the location of the gland as being near the ear, which indeed it is. **Parotid** will be used in this book.

The Colorado river toad *(Bufo alvarius)* is large, with the males having a snout-vent length (BL) of 8 to 15.6 cm and the females 8.7 to 17.8 cm (Figure 1).[4] Its body color has a greenish sheen.[5] The giant toad, *B. marinus,* (Figure 2), may reach a BL of 23 cm with dark brownish or grayish dorsal coloration.[5] The Columbian giant toad *(B. blombergi)* has a BL of 15 to 20 cm. Its back is a golden bronze, with sides of the body and the limbs black.[5] *B. regularis* has been reported as causing poisoning in dogs in Ethiopia.[7]

B. DISTRIBUTION

Amphibians are distributed throughout the world except in Arctic climates, but more species are found in the tropics than any other area.[4] Although most

TABLE 1
Classification of Poisonous Amphibians

Phylum — *Chordata*
Class — Amphibia
Order — Urodeles (Caudata or Urodela)
- Family — Salamandridae
  - Western American salamanders
  - California Newt — *Taricha torosa*
  - Rough-skinned newt — *T. granulosa*
  - Red-bellied newt — *T. rivularis*
  - Eastern U.S.
  - Spotted newt — *Notophthalmus viridescens*

Order — Anurans (Anura or Salientia)
- Family — Ranidae (true frogs)
  - *Dendrobates* spp. — poison arrow frogs
  - *Phyllobates* spp. — poison arrow frogs
- Family — Bufonidae (toads)
  - Colorado river toad — *Bufo alvarius*
  - Giant toad — *Bufo marinus*
  - Columbian giant toad — *Bufo blombergi*
  - Toads — *Bufo* spp.

FIGURE 1. Colorado river toad *Bufo alavarius*.

FIGURE 2. Marine toad *Bufo marinus*.

species of amphibians secrete substances that make them unpalatable, thus minimizing predation, only a few species produce substances capable of killing a dog. Few poisonous species inhabit areas accessible to domestic animals.

Poisonous salamanders are found in local populations in the northeastern U.S. and along the Pacific coast of the U.S. and Canada. Poison arrow frogs are neotropical species, inhabiting riverine areas in northern South America and Central America.[5] No poisonings have been reported in domestic animals.

All the toads in the genus *Bufo* produce poisonous secretions in the parotid gland and members of this genus are found all over the world. The species most often responsible for poisoning in dogs are native to North and South America, but have been introduced to numerous other countries to control insects, particularly in sugarcane growing areas.[8-10]

Bufo marinus inhabits South America, Central America, some of the islands in the Caribbean (Bermuda), and Texas, and has been introduced into Florida, Cuba, the Greater and Lesser Antilles, Hawaii, Fiji, Australia, the Philippine Islands, and the Marianas Islands.[4]

Bufo alvarius is adapted to a hot, desert environment but is restricted to the Imperial Valley of California and the valley of the lower Colorado River situated between Arizona and California and emptying into the Sea of Cortez in Mexico.[4] *Bufo blombergi* is found only in Columbia.

II. CONDITIONS OF POISONING

In the U.S., poisoning is seen primarily in the dog and only rarely in the cat.[11-13] It is not uncommon to see a naive puppy playing with a toad. The lethargic hopping of the awkward toad may attract young dogs at dusk.[14] If the puppy bites a toad, it will not likely do so again because the skin secretions are repugnant. Adult animals have usually had the experience, so poisoning is less likely.

A dog will usually grab a toad by the head and in so doing compresses the parotid glands, causing expression of the secretion.[13] The author was present when a blomberg toad squirted secretion into the eye of a person handling it. The handler experienced immediate irritation and rushed to a faucet to wash the material from the conjunctival sac. His conjunctival membranes remained hyperemic for a few hours, but he experienced no systemic effects.

In locations where *Bufo marinus* has been introduced to control insects, a population explosion has occurred and the toads have become pests. Native wild carnivores and omnivores in those areas may be poisoned as they prey on the exotic toad. As stated previously, giant toads have been introduced into almost every island or country that grows sugarcane. Such introductions have been a blessing in terms of insect control, but the large toads usually usurp the habitats of smaller native species, exploiting this new niche to the fullest. In Australia, these toads have invaded gardens and homes, bringing risk to household pets and perhaps being a hazard on the roads.[9,10]

There is no absolutely perfect deterrent to predation. Mammals, birds, and reptiles develop either mechanical means of coping with a poison or develop some degree of resistance to the effects of the poison. The North American raccoon, *Procyon lotor*, has learned to avoid ingestion of the skin of the Colorado river toad. The raccoon preys on the toad at spawning sites by pulling a toad out of the water, turning it over onto its back, and pulling open the abdomen to feed on the internal structures without mouthing the skin.[4]

Animals that normally prey on frogs and toads possess detoxification mechanisms to deal with the poison. As with most detoxification systems, a periodic stimulation of the system is required to keep the system functional.

Amphibian poisoning is seasonal and varies with the region. In northern temperate climates, spring and summer are the months of reproduction and increased density of toads.

Human beings have become seriously ill following exposure of the parotid gland secretion to scratches on the hands, or have died following ingestion of food contaminated with the poison.[15] In Hawaii, immature toads have been mistaken for frogs and have been eaten, with lethal consequences.[13] The effect on a human being's conjunctival membrane has already been described. If the secretion were not washed out quickly, it is likely that absorption and systemic reaction would result.

III. VENOM

The poisonous and medicinal qualities of toad skin and its special glands have been known for centuries. Ancient Chinese and Japanese cultures used dried toad extract as an expectorant, cardiac stimulant, and diuretic and included the extract in remedies for toothache and sinusitis.[14] Natives in restricted areas of South America (Columbia) and, to a lesser extent, Africa dipped dart tips to be used with a blowgun into a concoction prepared by heating a fresh frog skin to encourage the secretion to flow out of the skin.[14,16] However, this was never an extensive practice because most poison arrows and blowgun darts were prepared from poisonous plant mixtures.[16]

The secretion of the toad parotid gland is highly complex, but lacks the multitude of enzymes typical of snake venom. There are three major components.[8,15,17]

1. *Bufagins* have a cardioactive glycoside-like action that may culminate in ventricular fibrillation
2. *Bufotoxins* are the conjugation product of a specific bufagin with one molecule of suberyl arginine; the action is similar to that of bufagins
3. *Bufotenines* are organic bases containing an indole ring in the molecule; they have an oxytocic action, with little or no pressor action[11]

Other components of toad parotid secretion are:[8]

1. Epinephrine
2. Cholesterol
3. Ergosterol
4. Serotonin (5-hydroxytryptamine)

Poison dart frogs produce a highly potent poison from granular glands which are more diffusely distributed on the dorsum of the body than in toads.[16] The active principles of the secretion are alkaloids called "batrachotoxins" (*batrachos* = Greek for frog).[16]

Certain salamanders have granular glands that secrete a poison similar in action to that of toads,[1] and some newts (Table 1) contain tetrodotoxin in skin, blood, muscle, and eggs. Tetrodotoxin is one of the most potent marine biotoxins known and is found elsewhere in nature only in the puffer fish families *Tetraodontidae* and *Diodontidae*. Some of the 90 or more species of puffer fish are highly prized as food delicacies (fugu) in the Orient, *Tetradon* spp. and *Fugu* spp.[18] Special chefs must be trained to prepare the fish properly to prevent human poisoning. Nonetheless, eating fugu is a risk.

A. TOXICITY

The potency of the secretion produced by the parotid glands varies with the region and perhaps as a result of diet, climate, or unknown genetic factors. In

Hawaii, the mortality rate of untreated animals exposed to *Bufo marinus* is only 5%, while in Florida untreated exposure to the same species results in nearly 100% mortality. The mortality rate is also low in Texas.[12,14]

IV. CLINICAL SIGNS

Poison dart frogs are not involved in poisoning of animals except by injection. Signs of toad poisoning may occur within seconds to minutes of mouthing or ingestion of a toad and death may follow in as little as 15 min. The first sign is profuse salivation, followed by prostration, cardiac arrhythmia, pulmonary edema, convulsions, and death. Vomiting may occur in some dogs.[11,15,19,20] In Florida, the course of the disease is usually peracute, while elsewhere the disease may be less acute with spontaneous recovery possible.

The clinical signs of tetrodotoxin poisoning are those associated with a neurotoxin, with pupillary dilatation, weak respiratory effort, cardiac arrhythmia, and paralysis.[1] See Chapter 3 for more details.

V. DIAGNOSIS

A history of playing with or ingesting a toad is crucial to making an antemortem diagnosis. The signs are indistinguishable from acute, chlorinated hydrocarbon insecticide poisoning.[11] At necropsy, there may be pulmonary edema and, possibly, parts of the toad may be found in the stomach if the toad was ingested. Otherwise, no gross or microscopic lesions are helpful in making a diagnosis.

VI. TREATMENT

A. FIRST AID

If the dog has merely mouthed the toad, the poison may be quickly washed out of the mouth with water by the owner, using a garden hose.[11] This technique is not likely to save a dog that has ingested a toad, but it may remove some of the poison and decrease the total dose. Quick administration of an emetic may empty the stomach, but this should not be done once the dog has started convulsing.

B. MEDICAL

Historical medical treatment of toad poisoning consisted of the administration of pentobarbital sodium (Nembutal®) intravenously to control convulsions, and atropine to decrease salivation and control cardiac arrhythmias.[11,12] Other drugs included in older treatments included prednisone, antihistamines, calcium gluconate, and tranquilizers.[13,21,22] Unfortunately, the administration of some of these drugs may be counter-productive, if not contraindicated.

A combination of sublethal doses of ouabain (cardioactive glycoside) and epinephrine were shown to produced a lethal syndrome in rats similar to that seen in canine toad poisoning.[13,15] Subsequently, more emphasis has been given to specific therapy against the toxic components rather than to treating signs.

Toad poisoning is presumed to be caused by the cardioactive glycoside fraction of the bufotoxin complex with potentiation by endogenous catecholamines.[13,15] Propranolol HCl is a β-adrenergic blocking agent that is a specific antidote acting by antagonizing the glycosides' stimulatory effect on sympathetic nerves, thus blocking the release of endogenous catecholamines.[13] This action then limits the poisoning to essentially a nonlethal cardioactive glycoside-like effect, which the dog is able to handle.[13]

A recommended treatment once a tentative diagnosis of toad poisoning has been made is administration of propranolol HCl, (5 mg/kg, intravenously) as rapidly as possible, anesthesia with pentobarbital sodium, tracheal intubation, and rinsing the oral cavity with water should follow. Vital functions should be monitored every 10 min, preferably with an electrocardiograph (lead II).

The propranolol dose should be repeated in 20 min if necessary. The slow intravenous administration of potassium may be indicated if cardiac function does not return to normal within a few minutes.

VII. PROGNOSIS

Knowles[11] reports that almost 100% of dogs in Florida with a complete exposure will die if prompt veterinary care is not provided. There may be a slight difference in the toxicity of the poison in various parts of the U.S. because, in California and other places in the world, the syndrome is not so rapidly fulminating and the mortality rate of untreated dogs is not so high.

VIII. PREVENTION

When giant toads were first introduced into new areas, the mortality in the local dog population skyrocketed. As the dog population gained experience, dogs learned to leave the big toads alone. It is usually only the puppy or inexperienced dog that becomes a victim. During seasons of high toad density, puppies must be confined and watched carefully when being exercised. Public awareness programs have been effective in alerting owners of the danger and precautions to take. Governmental programs to control the toads have been uniformly unsuccessful.

REFERENCES

1. **Caras, R.,** The amphibians, in *Venomous Animals of the World,* Caras, R., Ed., Prentice-Hall, Englewood Cliffs, NJ, 1974, 133.
2. **Chen, K. K. and Chen, A. L.,** Notes on the poisonous secretions of twelve species of toads, *J. Pharm. Exp. Therapy,* 47, 285, 1933.
3. **Freytag, G. E.,** Urodeles and caecilians, in *Grzimek's Animal Life Encyclopedia,* Vol. 5, Grzimek, B., Ed., Van Nostrand Reinhold, New York, 1974, 309.
4. **Heusser, H. R.,** Higher Anurans, in *Grzimek's Animal Life Encyclopedia,* Vol. 5, Grzimek, B., Ed., Van Nostrand Reinhold, New York, 1974, 397.
5. **Lutz, B.,** Venomous toads and frogs, in *Venomous Animals and Their Venoms,* Vol. 2, Bucherl, W. and Buckley, E. E., Eds., Academic Press, Orlando, FL, 1971, 423.
6. **Porter, K. R.,** Dermal glands, in *Herpetology,* Porter, K. R., Ed., W. B. Saunders, Philadelphia, 1972, 23.
7. **Perry, B. D. and Bracegirdle, J. R.,** Toad poisoning in small animals, *Vet. Rec.,* 92, 289, 1973.
8. **Palumbo, N. E. and Perri, S. F.,** Toad poisoning, in *Current Veterinary Therapy, VIII. Small Animal Practice,* Kirk, R. W., Ed., W. B. Saunders, Philadelphia, 1983, 160.
9. **Covacevich, J. and Archer, M.,** Poisonous toads *Bufo marinus* in Queensland, *Mem. Qld. Mus.,* 17, 305, 1975.
10. **Seawright A. A.,** *Bufo marinus* (cane toad), in *Animal Health in Australia, Vol. 2. Chemical and Plant Poisons,* Seawright, A. A., Ed., Australian Government Publishing, Canberra, 1982, 255.
11. **Knowles, R. P.,** The poison toad and the canine, *Vet. Med. Small Animal Clin.,* Jan, 38, 1964.
12. **McCuistion, W. R.,** The toad and its repellent, *N. Am. Vet.,* August 1953, 569, 1953.
13. **Palumbo, N. E. and Perri, S.,** Toad poisoning, in *Current Veterinary Therapy, VI. Small Animal Practice,* Kirk, R. W., Ed., W. B. Saunders, Philadelphia, 1977, 173.
14. **Palumbo, N., Perri, S., and Read, G.,** Experimental induction and treatment of toad poisoning in the dog, *J. Am. Vet. Med. Assoc.,* 167, 1000, 1975.
15. **Otani, A., Palumbo, N., and Read, G.,** Pharmacodynamics and treatment of mammals poisoned by *Bufo marinus* toxin, *Am. J. Vet. Res.,* 30, 1865, 1969.
16. **Myers, C. W. and Daly, J. W.,** Dart-poison frogs, *Sci. Am.,* 248(2), 120, 1983.
17. **Meyer, K. and Linde, H.,** Collection of toad venoms and chemistry of toad venom steroids, in *Venomous Animals and Their Venoms, Vol. 2,* Bucherl, W. and Buckley, E. E., Eds., Academic Press, Orlando, FL, 1971, 521.
18. **Fowler, M. E.,** Zootoxins, in *Plant Poisoning and Biotoxins,* Fowler, M. E., Ed., Reprographics, University of California, Davis, 1987, 158.
19. **Fowler, M. E.,** Toxicities in exotic and zoo animals, *Vet. Clin. N. Am.,* 5, 685, 1975.
20. **Oehme, F. W., Brown, J. F., and Fowler, M. E.,** Toxins of animal origin, in *Casarett and Doull's Toxicology — The Basic Science of Poisons,* 2nd ed., Doull, J., Klassen C. D., and Amdur, M. O., Eds., Macmillan, New York, 1980, 557.
21. **Turbet, C. R.,** Poisoning of dogs by the giant toad *Bufo marinus:* its treatment with nembutal, *Austr. Vet. J.,* 14, 152, 1938.
22. **Russell, R. L.,** Toad poisoning, in *Current Veterinary Therapy, III. Small Animal Practice,* Kirk, R. W., Ed., W. B. Saunders, Philadelphia, 1968, 122.

Chapter 12

VENOMOUS LIZARDS

I. IDENTIFICATION AND BIOLOGY

A. IDENTIFICATION

Only 2 of the approximately 3000 species of lizards are venomous.[1,2] *Heloderma suspectum*, the gila monster (Figure 1) is comprised of two subspecies, *H. suspectum cinctum* and *H. suspectum suspectum*. There are three recognized subspecies of the Mexican beaded lizard (Figure 1): *Heloderma horridum exasperatum, H. horridum horridum*, and *H. horridum alvarezi*.[1-3]

Gila monsters and beaded lizards have a similar heavy body and tail and adults weigh from 1 to 2 kg with a total length of 250 to 800 mm. The head is large, with a bluntly rounded snout.[1] The tail is used for fat storage, so plumpness depends on the general body condition.[2] The tail stump does not regenerate should the tail become detached (autectomy).[1,4]

Teeth are long and recurved, with grooves on the rostral and caudal borders that direct the venom.[1,5] The largest teeth with the deepest grooves are those connected with the ducts of the venom gland.[5] The paired venom glands are multilobed, with separate ducts leading from each lobe to an orifice that discharges near the base of specific mandibular teeth.[1,5,6] Venom glands are modified labial glands located lateral to the rostral mandible (Figure 2).

The method of biting is to grasp and chew, thus facilitating injection of the venom.[5] Human victims report that the bite is powerful, and that the lizard may hold on tenaciously, requiring force to pry the jaws apart.[6]

Gila monster and beaded lizard diets consist primarily of the helpless infants of small mammals such as rats, mice, and rabbits, and equally helpless small birds. The eggs of birds and reptiles are also consumed.[2]

B. DISTRIBUTION

The gila monster is found in two populations in upper Sonoran desert habitat in Arizona, New Mexico, southern Utah, and Nevada, and the Mojave desert of California, with extension into Mexico in Sonora as far south as the Rio del Fuerte in Northern Sinaloa (Figure 3). Beaded lizards occupy a similar habitat on the western coast of Mexico from southern Sonora to Chiapas and on into northern Guatemala (Figure 3). The two populations overlap slightly in southern Sonora and Northern Sinaloa.[1]

II. CONDITIONS OF POISONING

Gila monsters are generally slow moving, inoffensive, nocturnal creatures. The vast majority of human bites occur when the victim is handling a captive specimen in an inappropriate manner.[5] Animal bites are extremely rare, but dogs or cats that harassed a gila monster have been bitten.

FIGURE 1. Mexican beaded lizard *Heloderma horridum*, top, and Gila monster *Heloderma suspectum*, bottom. (Photo by the author, courtesy Steinhart Aquarium, San Francisco, CA.)

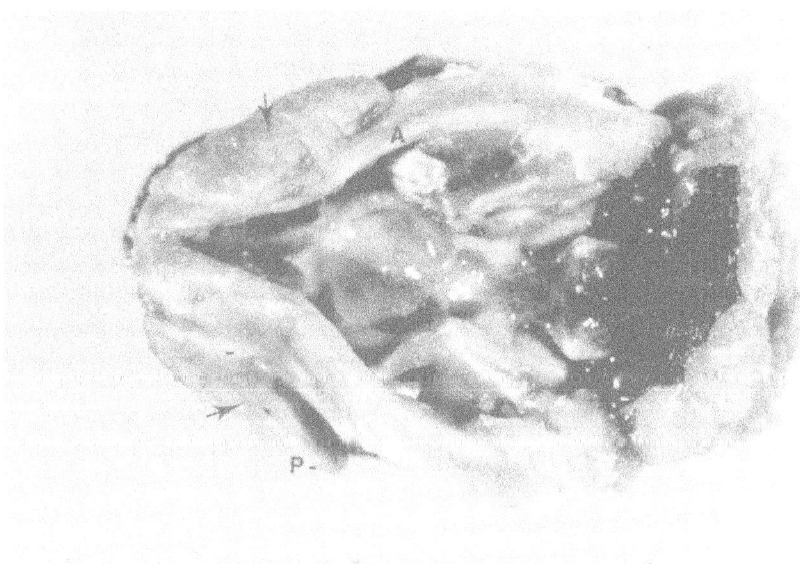

FIGURE 2. Location of the mandibular venom glands (arrows) of the gila monster, *Heloderma suspectum cinctum*. A, mandible.

FIGURE 3. Distribution of ▥ Gila monster *Heloderma suspectum cinctu* ▤ Gila monster *H. s. suspectum* ▦ Beaded lizard *H. horridum*. (Adapted from Campbell, J. A. and Lamar, W. W., *The Venomous Reptiles of Latin America*, Campbell, J. A. and Lamar, W. W., Eds., Comstock Publishing, Ithaca, New York, 1989, 85.)

The desert has become a popular recreation ground for off-road vehicle enthusiasts and campers, often accompanied by pets. City pets that are unaccustomed to strange animals may investigate too closely.

Gila monsters and beaded lizards are apparently one of the few venomous species that have evolved a lethal venom and a rather primitive envenomation apparatus that is used strictly for defense. None of the lizards' prey require immobilization prior to eating and the venom has no characteristics that enhance digestion.[2,6]

III. VENOM

A. COMPOSITION

The venom of the gila monster and beaded lizard are similar in composition and effect on vertebrate systems. The pharmacological effects of the crude venom and the nine purified components isolated from gila monster venom have been studied extensively.[3] The venom is antigenically unrelated to those of snakes. The major lethal factor has been given the name of gilatoxin and is devoid of proteolytic, hemorrhagic, or hemolytic activity.[3,7] Enzymatic action includes phospholipase A_2 and hyaluronidase.

B. TOXICITY

Heloderma venom has been known for its lethal qualities for over 100 years. Reports of investigations utilizing dogs state that 5 to 6 drops of fresh venom administered intravenously were lethal within 15 min.[8] The subcutaneous LD_{50} for a mouse is 0.82 to 4.0 mg/kg and a rat, 14.0 mg/kg.[3,8]

IV. CLINICAL SIGNS

Little information has been published in the literature concerning gila monster bites in dogs and cats. Dogs have been used as experimental subjects and signs noted during poisoning seem to be similar to those experienced by human bite victims. The wounds consist of multiple puncture marks that may become obliterated by edema that may develop within 15 min or gradually over 4 to 8 h. Fractured teeth may be embedded in the wound. Intense, localized pain may be indicated by the victim biting or scratching at the swelling.

Heloderma venom has little or no necrotoxic effect. Hypotension, tachycardia, and dyspnea have been reported in experimental subjects.[3] Hypovolemic shock may be observed.

V. DIAGNOSIS

Without a history of exposure to a gila monster, diagnosis may be difficult. A presumptive diagnosis may be made if the typical bite wound is visible and accompanied by edema and evidence of pain. A differential diagnosis should include insect and arachnid stings, foreign body penetration, contusion, and snakebite. No laboratory findings are definitive for gila monster bites.[6]

VI. TREATMENT

A. FIRST AID

Veterinarians may receive emergency calls from clients in remote locations, soliciting recommendations for administration of first aid. If the bite has been witnessed, the first problem may be to disengage the lizard from the pet. Chewing activity enhances venom deposition in the wounds; thus, the shorter the time the lizard is attached, the better the chances for minimal envenomation. If the lizard is determinedly hanging on, a metal bar or other sturdy lever will be needed to pry the jaws open. Pulling on the lizard will convert puncture wounds into multiple lacerations.

All removal efforts should be carried out safely by first grasping the lizard at the neck. A flame from a match or cigarette lighter applied to the bottom jaw may encourage the lizard to disengage. The lizard may be killed by incising the jugular vein with a knife or by a blow to the head with a rock; however, the latter method may drive the teeth deeper into the wound.[6]

The bitten limb should not be packed in ice nor should any type of constricting band be applied around the limb.[6]

B. HOSPITAL MANAGEMENT

Any known gila monster bite victim should be moved to the nearest veterinarian for assessment and monitoring. Death from such a bite is rare in both animals and humans, primarily because modern medical practice allows management of the patient as clinical signs arise.

The bite site should be cleansed thoroughly and individual puncture wounds irrigated with 2% lidocaine delivered from a blunted 26-gauge needle. Pain may be ameliorated by appropriate analgesic drugs. Diazepam, 0.1 to 0.5 mg/kg i.m., will quiet an excited victim and provide relief from pain as well.

The patient should be monitored for signs of hypotension and shock. It may be desirable to place an intravenous catheter, in case fluid therapy becomes necessary. No known laboratory findings will aid in diagnosis or can be used to assess progression of the course of a *Heloderma* bite, but having baseline information available may aid in differential diagnosis of other conditions.

Any bite wound is potentially septic. The veterinarian must determine the desirability of antibiotic therapy or the use of antitetanus agents (antitoxin/toxoid).[6]

VII. PROGNOSIS

Death from a gila monster bite in a dog or cat is rare, especially with appropriate medical care followed by close observation for 3 to 5 d.

VIII. PREVENTION

Responsible animal ownership minimizes the risk of envenomation. Pets should not be allowed to roam freely outside the home and yard environment.

REFERENCES

1. **Campbell, J. A. and Lamar, W. W.,** Gila monster and beaded lizard, in *The Venomous Reptiles of Latin America,* Campbell, J. A. and Lamar, W. W., Eds., Comstock Publishing, Ithaca, NY, 1989, 85.
2. **Caras, R.,** The venomous lizards, in *Venomous Animals of the World,* Caras, R., Ed., Prentice-Hall, Englewood Cliffs, NJ, 1974, 145.
3. **Tu, A. T.,** A lizard venom: Gila monster *Heloderma,* in *Handbook of Natural Toxins, Vol. 5. Reptile Venoms and Toxins,* Tu, A. T., Ed., Marcel Dekker, New York, 1991, 755.

4. **Tinkham, E. R.,** The biology of the gila monster, in *Venomous Animals and Their Venoms,* Vol. 2, Bucherl, W. and Buckley, E. E., Eds., Academic Press, Orlando, FL, 1971, 387.
5. **Kochva, J. E.,** Venom gland of *Heloderma,* in *Biology of the Reptilia,* Vol. 8, Gans, C., Ed., Academic Press, Orlando, FL, 1978, 80.
6. **Russell, F. E. and Bogert, C. M.,** Gila monster: its biology, venom and bite — A review, *Toxicon,* 19, 341, 1981.
7. **Tinkham, E. R.,** The venom of the gila monster, in *Venomous Animals and Their Venoms,* Vol. 2, Bucherl, W. and Buckley, E. E., Eds., Academic Press, Orlando, FL, 1971, 415.
8. **Mebs, D.,** Pharmacology of the venom of the *Heloderma,* in *Biology of the Reptilia,* Vol. 8, Gans, C., Ed., Academic Press, Orlando, FL, 1978, 438.

Chapter 13

SNAKE VENOMS

I. INTRODUCTION

Snake venoms contain complex mixtures of various organic materials (proteins, peptides, amino acids, and other compounds).[1] Snake venoms have been of interest to scientists and laypersons alike for thousands of years, but only during the last 3 decades has biochemical technology allowed isolation and precise identification of the toxins (venins, constituents) in venoms. Isolation of specific toxins has enabled scientists to study the role of each toxin in the overall action of the venom.

A basic understanding of venoms is necessary to understand the rationale for certain treatment regimens. Hundreds of research papers and monographs have been written on the subject of snake venoms by protein biochemists, toxinologists, toxicologists, physiologists, pharmacologists, immunologists, and medical scientists. This short discussion is meant only to provide an introduction and overview of this rapidly expanding area of venom research. More detail may be obtained from reviews of the subject.[2-8]

Though much of snake venin research has been directed at finding solutions for problems associated with envenomation of humans, most research studies have employed experimental animals or isolated tissue. Different species may react differently to venins; still, much can be learned about envenomation in general by understanding the effects of snake toxins on mice, rabbits, sheep, and other animals.

II. COMPOSITION

The primary toxins of snake venoms are proteins and peptides. New toxins are being described almost monthly. Table 1 lists known toxins according to biochemical components, while Table 2 classifies toxins according to their primary effect on mammalian tissues or organ systems. Toxins are usually named according to the snake venom from which the toxin was first isolated, or the primary pharmacologic effect on mammalian victims.[9] Many toxins have not yet been precisely identified and are currently designated only as fraction x, y, or z from a given source. Snake venoms may vary, even between closely related species.[10,11] Toxin nomenclature is anything but standardized, and multiple names may exist for the same toxin studied in various areas of the world. The author has followed the nomenclature found in Tu's *Reptile Venoms and Toxins*.[7]

Enzymes are the major toxic components of snake venoms.[12] Phospholipase A_2 activity (i.e., hydrolytic breakdown of membrane phospholipids) is basic to the action of many of the toxins in snake venoms. Pharmacologically, phospholipases A_2 may have neurotoxic, cytotoxic, and anticoagulant activities.

TABLE 1
An Outline of the Chemical Components of Snake Toxins

Chemically, a typical venom is 90% water with 5–15 enzymes and numerous proteins, peptides, and amino acids. A few nonprotein compounds have been identified, but are of little consequence in toxicity.

Organic Proteins

Enzymes (A total of 26 enzymes have been isolated from various snake venoms — most are hydrolytic)

- Phospholipases A_2 (Lecithinase) — containing:
  - β-Neurotoxins
    - Single-chain β-neurotoxins
      - Ammodytoxin A
      - Notexin
      - Caudoxin
      - Agkistrodotoxin
      - Pseudexin
    - Multichain β-neurotoxins
      - Ceruleotoxin
      - Crotoxin
      - Mojave toxin
      - Taipoxin
      - Textilotoxin
    - β-Bungarotoxin structure
      - β-Bungarotoxin
  - α-Neurotoxins
- Phospholipases B and C
- Acetylcholinesterase
- Adenosinetriphosphatase
- 5′-Nucleotidase
- Deoxyribonuclease
- Phosphodiesterase
- Hyaluronidase
- Proteases
- Lactate dehydrogenase
- Arginine ester hydrolase
- Collagenase
- Phosphomonoesterase
- RNase
- DNase
- L Amino acid oxidase (not hydrolytic)

Miscellaneous components

- Amino acids
- Nucleotides
- Carbohydrates (glycoproteins)
- Lipids (phospholipids)
- Biogenic amines
  - Bradykinin, histamine, serotonin

TABLE 2
An Outline of Snake Toxins by Toxic Action

- Presynaptic neurotoxins
  - β-Neurotoxins
    - Ammodytoxin A
    - Notexin
    - Caudoxin
    - Agkistrodotoxin
    - Ceruleotoxin
    - Crotoxin
    - Mojave toxin
    - Vipera toxin
    - Taipoxin
    - Textilotoxin
    - β-Bungarotoxin
  - Dendrotoxins
  - Antiacetylcholinesterase toxins
- Post-synaptic neurotoxins
  - α-neurotoxins
- Cardiotoxins — A number of snake venoms contain cardioactive toxins that are independent of phospholipase A_2 activity
  - Direct lytic factor
  - Cytotoxin
  - Cobramine
  - Cardiotoxin
- Complement-activating protein of cobra venom
- Thrombinlike enzymes
  - Ancrod
  - Batroxobin
  - Catroxobin
  - Crotalase
  - Flavoxobin
  - Gabonase
- Platelet inhibition and stimulation
  - Botrocetin
- Coagulation and anticoagulation activity
  - Snake venom proteases
  - Thrombocytin
  - Russell's viper venom-V
- Cytotoxic (necrotoxic) action
  - Cytotoxins
  - Proteases
  - Myotoxins
  - Phospholipase toxins
  - Hemorrhagic toxins
- Nephrotoxins

As a presynaptic neurotoxin (β-neurotoxin), phospholipase A_2 acts on peripheral axon terminal membranes in such a way as to decrease the release of acetylcholine into the neuromuscular junction. Post-synaptic neurotoxins (α-neurotoxins) bind to acetylcholine receptors on post-synaptic membranes to prevent uptake of the neural transmitter.[13] Both pre- and post-synaptic neurotoxins effectively paralyze skeletal muscles, including the diaphragm, thereby immobilizing and asphyxiating the victim.

Cytotoxic phospholipases A_2 form lysolipids that act like detergents in lysing cells, causing necrosis of muscle tissue (myotoxins). As anticoagulants, phospholipases A_2 eliminate the procoagulant phospholipids before blood clotting factors can be activated and begin amplification of the clotting cascade.[13] Descriptions of more detailed biochemical reactions are beyond the scope of this book.

III. NEUROTOXINS

A. β-NEUROTOXINS

Many β-neurotoxins have been recognized in venoms of elapid snakes, but they are also found in crotalid and viperid venoms (Table 2). These toxins effectively block nerve impulse transmission at the presynaptic location and paralyze the victim.[14-16]

B. α-NEUROTOXINS

Some of the most potent toxins known are α-neurotoxins, usually found in elapid snake venoms. Although α-neurotoxins act at the post-synaptic location by binding to nicotinic acetylcholine receptors, the effect is again a paralysis of the peripheral nervous system.[17,18]

IV. CARDIOTOXINS

Cardiotoxins may act independently or synergistically with phospholipase A_2, particularly in causing a lytic effect on erythrocytes. Cardiotoxins cause prolonged muscular contracture with subsequent loss of the ability of cardiac muscle to contract in response to normal stimuli.[17] Russell[19] lists the following biological activities of cardiotoxin: blockade of neuromuscular transmission, blockade of axonal conduction, membrane depolarization, anticholinesterase action, local tissue action, hemolysis, cytotoxic action, skeletal muscle contracture, smooth muscle contracture, and cardiac arrest. The biochemical method of action is unknown.

V. COMPLEMENT-ACTIVATING PROTEIN

Complement is not a single entity, but a complex proteinaceous enzyme system or cascade facilitating antibody-mediated reactions and induction of major inflammatory pathways. A cobra venom factor (CVF) that had a lytic effect on erythrocytes was identified at the end of the 19th century. Using modern biochemical technology, CVF has been purified and tested for physiologic action. Administered intravenously, CVF has little toxic effect, but its action at a bite site triggers activation of complement, releasing anaphylatoxins which in turn cause an increase in vascular permeability. Increased vascular permeability aids in the absorption of toxins into the bloodstream.[20]

VI. PLATELET-ACTIVE PRINCIPLES

Venom platelet-active principles include proteins, peptides, and glycoproteins that may either stimulate or inhibit platelet action. Agents that stimulate platelets include phospholipases A_2, serine proteases, and glycoproteins. Inhibitory toxins also include phospholipase A_2 activity and glycoproteins, plus peptides and fibrinogenases. Platelet-active toxins are more common in the venoms of viperids and crotalids than in elapid snake venoms.[21] These actions may contribute to thrombocytopenia and/or hemorrhage.

VII. BLOOD CLOTTING FACTORS

Thrombin-like venom enzymes induce the clotting of fibrinogen (Table 2).[22,23] These enzymes are primarily found in crotalid venoms, less so in viperid venoms. Certain toxins (e.g., Russell's viper venom-V and thrombocytin) activate Factor V in the blood coagulation cascade.[24] Other snake venoms activate protein-C in the coagulation sequence.[25] Certain factors in snake venoms may inhibit or stimulate the clotting mechanism, contributing to disseminated intravascular clotting (DIC) or hemorrhage.

VIII. CYTOTOXIC ACTION

Envenomation by crotalid snakes usually results in a marked local response characterized by erythema, hemorrhage, edema, and necrosis. Response at the bite site is visible, but identical action may take place in the brain, lung, kidney, heart, or liver, quickly in severely affected victims, or as the effects of envenomation progress.[26] This progressive effect should be kept in mind when monitoring patients or contemplating the use of antivenin. It is not true that antivenin has no inhibitory effect on local reactions to snakebite.

Hemorrhagic toxins attach to endothelial cells of blood vessels causing swelling and rupture of membranes, resulting in hemorrhage.[27] Hemorrhagic toxins are proteolytic in action.[26] Myonecrosis may be the consequence of the action of myotoxins on the sarcoplasma of muscle cells.[26]

IX. NEPHROTOXINS

Although specific nephrotoxins have not been identified, there is ample evidence that nephrotoxicity occurs in snakebite, particularly from the venom of such viperid snakes as Russell's viper. The cytotoxic effects of crotalid venoms may also cause renal lesions. The range of renal lesions includes acute tubular necrosis, acute cortical necrosis, glomerulonephritis, and acute interstitial nephritis. Renal failure may develop late in the course of an envenomation, another reason for continual monitoring of the patient for signs of parenchymal organ dysfunction. Signs of renal involvement include hemoglobinuria, myoglobinuria, disseminated intravascular coagulation, hypotension, and hemorrhage.[28]

Aside from the importance of their poisonous effects, snake venoms have proven to be valuable tools in the study of nerve conduction, blood coagulation, tumor growth, and enzymology.[29]

X. VENOM TOXICITY

The primary method of determining the toxicity of snake venoms and individual toxins is to establish the lethal dose in 50% of the experimental animals of a given species into which the venom is injected (LD_{50}) (see Table 3). The mouse, *Mus musculus,* is the usual subject. A perusal of the literature quickly reveals great variation in the values reported by various authors. The method of collection, time of collection, and how the venom is handled following collection may have a significant bearing on the potency of the venom.

Toxicity as established in the laboratory may differ markedly from the effects of an actual bite. Table 4 lists variables that may modify the ultimate effect of a venom. While some data on the toxicity of venoms for laboratory species is available, little factual data has been collected on toxicity for domestic and wild animals.

XI. VENOM APPARATUS

The primary function of snake venom is to procure food. Most venomous snakes stalk prey, strike, and release, waiting for immobilization to occur before grasping and ingesting the animal. Venom also serves a beginning digestive function. Secondarily, venom may be used defensively.

A. ELAPID SNAKES

The elapids have front-fixed fangs (proteroglyphous) (see Figure 1). The length of the fangs varies from species to species, with the Sea snake and North American coral snakes having short fangs. Fangs are partially covered by a membrane that is pushed away at the time of the bite. Elapid fangs are either grooved or have a closed groove that functions in the same manner as a hollow fang. The venom duct empties at the base and bathes the fang. Although there is species variation, the venom gland (homologous with the mammalian parotid salivary gland) generally lies ventral to the eye, dorsal to the upper lip,

TABLE 3
Relative Toxicity of Whole Snake Venom in Mice

Scientific name	Common name	Dose (LD_{50}) µg/kg i.v.
Colubridae		
Dispholidus typus	Boomslang	71.0
Crotalidae		
Agkistrodon acutus	Habu	380.0
A. contortrix mokasen	American copperhead	2711.0
A. piscivorus	Water moccasin	4000.0
Lachesis mutus	Bushmaster	4500.0
Bothrops atrox	Common lancehead Fer-de-lance	1400.0
Bothrops asper	Terciopelo Fer-de-lance	1244.0
Bothrops jararacussu	Jararacussu	460.0
Crotalus adamanteus	Eastern diamondback rattlesnake	2000.0
C. atrox	Western diamondback rattlesnake	3600.0
C. durissus terrificus	South American rattlesnake, Neotropical rattlesnake	1244.0
C. scutulatus scutulatus	Mojave (green) rattlesnake	100–200.0
C. viridis helleri	Northern Pacific rattlesnake	844.0
Sistrurus milarius	Pygmy rattlesnake	12,590.0
Trimeresurus flavoviridis	Okinawan habu	4–210.0
Trimeresurus gamineus	Indian tree viper	400.0
Viperidae		
Bitis arietans	African puff adder	2350.0
Bitis gabonica	Gaboon viper (adder)	1000–3000
Bitis nasicornis	Rhinoceros viper	8600.0
Echis carinatus	Saw-scaled viper	1200.0
Vipera aspis	European asp	1000.0
Vipera xanthina palaestinae	Palestine viper	300.0
Vipera russelli	Russell's viper	80–2000.0
Elapidae		
Austrelaps superbus	Australian copperhead	500.0
Bungarus fasciatus	Krait	1200–4000
Dendroaspis polypasis	Black mamba	250–300
Hemachatus hemachatus	Rinkals	1000–2000
Micrurus fulvius fulvius	Eastern coral snake	378.0
Naja haje	Egyptian cobra	420–1000
Naja hannah	King cobra	1300–1600
Naja naja	Asian cobra	130.0
Naja nigricollis	Spitting cobra	2600.0
Notechis scutulatus	Tiger snake	256.0

TABLE 3 (continued)
Relative Toxicity of Whole Snake Venom in Mice

Scientific name	Common name	Dose (LD_{50}) µg/kg i.v.
Oxyuranus scutulatus	Taipan	22.0
Pseudechis australis	Mulga, King brown	300.0
Pseudechis colletti	Collett's snake	840.0
Pseudechis porphriacus	Red-bellied black snake	540.0
Pseudonaja nuchalis	Gwardar, Western brown	473.0
Pseudonaja textillis	Eastern brown	10.0
Hydrophiidae		
Hydrophis spp.	Sea snakes	24–350.0

From Khole, V., *Handbook of Natural Toxins, Vol. 5. Reptile Venoms and Toxins,* Tu, A. T., Ed., Marcel Dekker, New York, 1991, 405.

TABLE 4
Factors Modifying Toxic Effects of Snake Venoms

Victim factors
- Age
- Size — weight
- Sex — female mice more susceptible than males[a]
- General condition
- Species — animals vary in response to venom
- Tissue injected — fat less absorptive

Snake factors
- Number of fangs inserted
- Length of fangs
- Single or multiple bites
- Quantity of venom injected
  - Recent feeding
  - Size of the snake
  - The snake may not inject venom (dry bite)
- Age of the snake — older snakes may have more potent venom
- Potency of the venom
- Temperament of the snake

Environmental factor
- Ambient temperature — increasing the temperature may decrease the lethality of a venom[a]

[a] See Russell, F. E., *Snake Venom Poisoning,* Scholium International, Great Neck, New York, 1983.

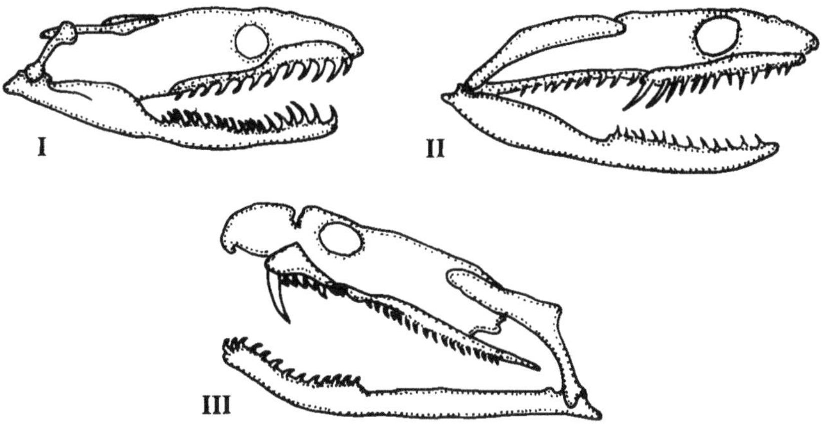

FIGURE 1. Diagrams of teeth and fangs: I, nonvenomous colubrid snake; II, rear-fanged colubrid snake; and III, front-fanged elapid snake.

and extends caudally to the commissure of the mouth. Usually, a single duct connects the gland to the base of the fang.

Smaller elapid snakes bite and continue to chew while holding on to the victim. Larger snakes may bite repeatedly, especially in a defensive situation.

When agitated, cobras rear up to face danger. Striking distance is measured by the height of the head above the ground because the snake strikes forward and downward, with the fulcrum being the point of contact of the snake with the ground. Cobras are relatively slow in striking speed, which enables the mongoose, *Herpestes* spp., family *Viverridae*, to dodge and ultimately grasp the snake.

A few elapid snakes (e.g., spitting cobras) have evolved a novel means of defense, utilizing venom and specialized fangs that have a distal orifice directed at a right angle to the fang. When threatened, the snake elevates the head and sprays two jets of venom toward the eyes of the animal. The venom may be projected accurately a distance of 3 to 4 m. The venom causes an immediate conjunctivitis and keratitis that results in temporary blindness. Spitting cobras immobilize prey by biting in a normal cobra manner.

B. VIPERID AND CROTALID SNAKES

Viperid and crotalid snakes have elongated, hollow, front fangs (solenoglyphous) that lie folded on the roof of the mouth, covered by a membranous sheath, when not in use (Figures 2 and 3). Location of the venom gland is similar to that of elapid snakes.[30] A primary, convoluted duct connects the venom gland to an accessory gland and a secondary duct continues on, emptying into the base of the sheath opposite the proximal orifice of the fang. Venom then travels through the venom canal to the distal orifice on the rostral aspect near the tip of the fang.[19]

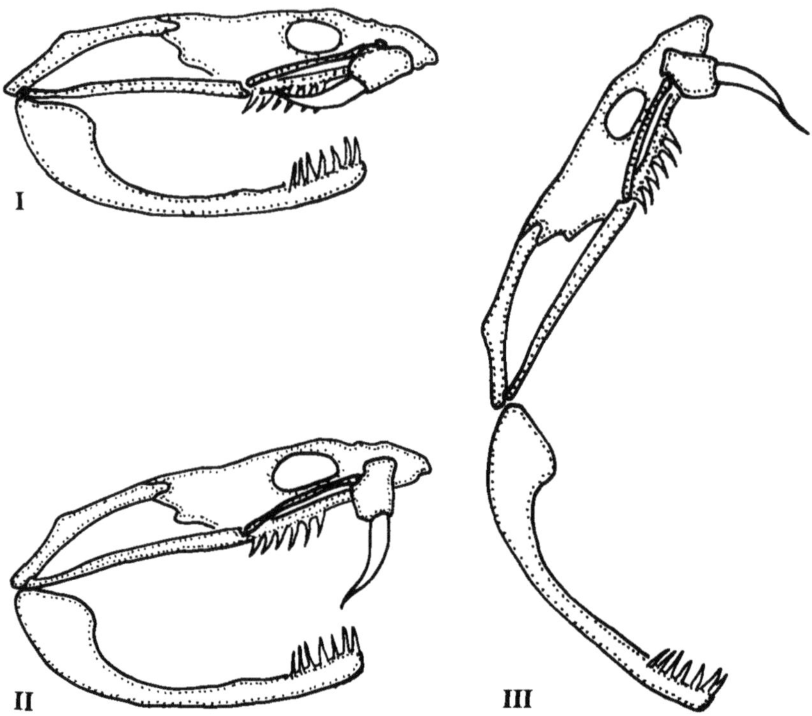

FIGURE 2. Diagrams of crotalid or viperid fang and skull anatomy: I, fang folded to the roof of the mouth; II, fang extended; and III, mouth open for strike.

Fangs are shed periodically (6 to 10 weeks) and reserve fangs may be present, confusing the expected picture of two fang marks at a bite site.[19] Conversely, only one functional fang may be present at a given time or the nature of the strike may produce a single fang injection. Viperid, elapid, and crotalid snakes also have rows of mandibular, maxillary, and palatine teeth that may puncture the skin.

Vipers and pit vipers may envenomate from various positions. If grasped or stepped upon, the snake immediately twists to bite the hand or limb. This bite would be similar to an elapid bite. Venom may be injected by one or both fangs, at a depth determined by the length of the fang.

A strike is a different maneuver; the snake positions itself into "S"-shaped folds, with the head slightly elevated. As the snake begins the strike, the mouth is opened to approximately a 160° angle, the fangs are erected to a 90° angle with the upper jaw, and the head is thrust forward toward the victim (Figure 3). The bite resulting from a strike of a viperid or crotalid snake is more like a hypodermic injection than a bite. The forward thrust of the curved fang on penetration of the skin may initiate closure of the fang, so that even a long-fanged snake may make a subcutaneous rather than a deep, intramuscular injection. The depth of the injections and the amount of venom injected vary with the size and aggressiveness of the snake.

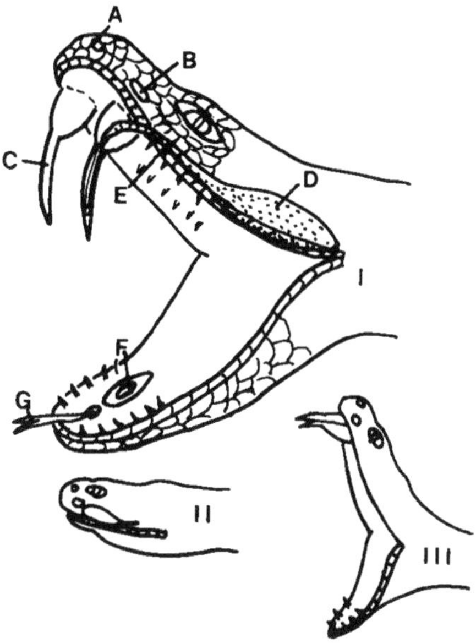

FIGURE 3. Diagrams of the venom apparatus of a crotalid or viperid snake. I: fangs extended; A, nostril; B, labial (heat sensitive) pit in crotalid; C, fang; D, venom gland; E, venom duct; F, glottis; and G, tongue. II: fangs folded to the roof of the mouth. III: mouth open in strike position.

The maximum striking distance is approximately two thirds the body length of the snake. Viperids and crotalids cannot propel the entire body through the air.

C. COLUBRID SNAKES

Venomous snakes in the family *Colubridae* have one to three elongated, fixed maxillary teeth (opisthioglyphous) at the caudal aspect of the dental arcade (rear-fanged) (Figure 1). Most snakes in this family are not considered dangerous to humans or animals, but human fatalities have been associated with the bite of the boomslang, *Dispholidus typus,* and illness has been produced by bites of the mangrove, *Boiga dendrophila,* and the California lyre snake, *Trimorphodon vandenburghi.*

Colubrid venom is produced in a special gland called "Duvernoy's gland", the ducts of which empty into the mouth at the base of the fangs. The fangs are grooved on the rostral border and the venom is carried into the wound by capillary action during the chewing associated with the bites of these snakes.

XII. RESISTANCE TO SNAKE VENOM

Species response to snake envenomation is marked. Some of the factors involved are unknown, but immunity to the proteinaceous compounds in the venom may be an important protective mechanism. It has been assumed since antiquity that venomous snakes may be immune to their own venom. Pliny the

Elder said, "The sting of the serpent is not aimed at the serpent."[31] Modern technologies have established that neutralizing antibodies in the serum of the eastern diamondback rattlesnake *(Crotalus adamanteus)* protect the snake from accidental, self-inflicted envenomation.[32] Serum from the rattlesnake is more effective than commercial antivenin for protecting mice from the effects of rattlesnake venom. The same is true of water moccasin *(Agkistrodon piscivorus)* serum against water moccasin venom.

Similar results have been obtained from studies of the resistance of viperid and elapid snakes and the gila monster, *Heloderma suspectum*, to their own venom.[32-35] It has been stated that the Australian tiger snake, *Notechis scutatus*, is 108,000 times more resistant to its own venom than is the guinea pig, *Cavia porcellus*.[32] Such nonvenomous snakes as the king snake, *Lampropeltis getulus*, that prey upon other snakes and include crotalids in their diet are also resistant to crotalid venom.[32-36]

Knowledge of resistance to self-injected venom is important to the manager of captive collections of venomous snakes. It is not uncommon for a hyperexcited snake to bite itself during restraint procedures necessary for medical care. However, protection is not absolute, but is dependent upon the injected dose of venom.

Great diversity exists in mammalian resistance to snake venoms. Although the mongoose is highly adept at avoiding the slow strike of a cobra, it has also long been reputed to be resistant to cobra venom, and this has now been demonstrated in the laboratory.[37] Nevertheless, a large venom dose may exceed the natural resistance and kill the mongoose if it makes a mistake and grasps a cobra too far caudal on the neck, allowing the snake to turn and bite.

Other small carnivorous mammals that include snakes in their diet may also have natural resistance to the venom of snakes in their habitat. The Virginia opossum, *Didelphis virginiana*, is relatively resistant to crotalid venom.[38] Likewise, the European hedgehog, *Erinaceus europaeus*, is capable of neutralizing some of the hemorrhagic factors in viper venom.[39]

Domestic and feral pigs *(Sus scrofa)* are also reputed to be resistant to crotalid venom, but the protection is provided by resistance of the thick skin to fang penetration and the layer of subcutaneous fat that retards absorption of the venom into the bloodstream. Intramuscular injection of venom may be lethal to the pig.[37]

Birds apparently have no immunity against snake venoms, even though raptorial birds may attack and kill venomous snakes. Feathers may prevent the fangs from penetrating the skin.[37]

Some humans have claimed immunity to snakebite, particularly members of snake-worshipping cults and individuals who have been repeatedly bitten by venomous snakes. Solid evidence for long-term immunity is lacking.[31,36] An Australian herpetologist allowed himself to be hyperimmunized and, at the end of a year of repeated injections, was receiving eight times the estimated lethal dose of tiger snake venom. At that time, his serum titer response was high, but a year later the titer had dropped and was likely unprotective.[37] Attempts to vaccinate humans in high-risk areas have been less than satisfactory.[37] In addition, individuals who have been repeatedly bitten by venomous snakes face the risk of anaphylaxis resulting from acquired sensitivity to the foreign proteins in the venom.[31]

REFERENCES

1. **Devi, A.,** The protein and nonprotein constituents of snake venoms, in *Venomous Animals and Their Venoms, Vol. 1. Venomous Vertebrates,* Bucherl, W., Buckley, E. E., and Deulofeu, V., Eds., Academic Press, Orlando, FL, 1968. 119.
2. **Bucherl, W., Buckley, E. E., and Deulofeu, V.,** *Venomous Animals and Their Venoms,* Academic Press, Orlando, FL, 1968.
3. **Rosenfeld, G. and Kelen, E. M. A.,** *Bibliography of Animal Venoms, Envenomations, and Treatments (period 1500–1968),* Sao Paulo, Brazil, 1969.
4. **Russell, F. E. and Saunders, P. R.,** *Animal Toxins,* Pergamon Press, New York, 1966.
5. **Russell, F. E. and Scharffenberg, R. S.,** *Bibliography of Snake Venoms and Venomous Snakes,* Bibliographic Association, West Covina, CA, 1964.
6. **Russell, F. E.,** *Poisonous Snakes of the World,* NAVMED, P-5099, Department of Navy, Bureau of Medicine and Surgery, U.S. Government Printing Office, Washington, D.C., 1965.
7. **Tu, A. T., Ed.,** *Handbook of Natural Toxins, Vol. 5. Reptile Venoms and Toxins,* Marcel Dekker, New York, 1991.
8. **Vick, J. A., Ciuchata, H. P., and Manthei, J. H.,** Pathophysiological studies of ten snake venoms, in *Animal Toxins,* Russell, F. E. and Saunders, P. R., Eds., Pergamon Press, Oxford, England, 1967, 269.
9. **Tu, A. T.,** Snake venoms, in *Venoms: Chemistry and Molecular Biology,* John Wiley & Sons, New York, 1977, 21.
10. **Tu, A. T., Toom, P. M., and Murdock, D. S.,** Chemical differences in the venoms of genetically different snakes, in *Animal Toxins,* Russell, F. E. and Saunders, P. R., Eds., Pergamon Press, Oxford, England, 1967, 351.
11. **Yang, C. C.,** Chemistry and evolution of toxins in snake venoms, *Toxicon,* 12, 1, 1974.
12. **Sarkar, N. K. and Devi, A.,** Enzymes in snake venoms, in *Venomous Animals and Their Venoms, Vol. 1. Venomous Vertebrates,* Bucherl, W., Buckley, E. E., and Deulofeu, V., Eds., Academic Press, Orlando, FL, 1968, 167.
13. **Davidson, F. F. and Dennis, E. A.,** Structure, function and mode of action of snake venom and other phospholipases A_2, in *Handbook of Natural Toxins, Vol. 5. Reptile Venoms and Toxins,* Tu, A. T., Ed., Marcel Dekker, New York, 1991, 107.
14. **Hawgood, B. and Bon, C.,** Snake venom presynaptic toxins, in *Handbook of Natural Toxins, Vol. 5. Reptile Venoms and Toxins,* Tu, A. T., Ed., Marcel Dekker, New York, 1991, 3.
15. **Hendon, R. A. and Bieber, A. L.,** Presynaptic toxins from rattlesnake venoms, in *Rattlesnake Venoms,* Tu, A. T., Ed., Marcel Dekker, New York, 1982, 211.
16. **Su, C., Chang, C. C.. and Lee, C. Y.,** Pharmacological properties of the neurotoxin of cobra venom, in *Animal Toxins,* Russell, F. E. and Saunders, P. R., Eds., Pergamon Press, Oxford, England, 1967, 259.
17. **Harvey, A. L.,** Cardiotoxins from cobra venoms, in *Handbook of Natural Toxins, Vol. 5. Reptile Venoms and Toxins,* Tu, A. T., Ed., Marcel Dekker, New York, 1991, 85.
18. **Atassi, M. Z.,** Postsynaptic-neurotoxin-acetylcholine receptor interaction and the binding sites on the two molecules, in *Handbook of Natural Toxins, Vol. 5. Reptile Venoms and Toxins,* Tu, A. T., Ed., Marcel Dekker, New York, 1991, 53.

19. **Russell, F. E.**, *Snake Venom Poisoning*, Scholium International, Great Neck, New York, 1983.
20. **Vogel, C.-W.**, Cobra venom factor: the complement-activating protein of cobra venom, in *Handbook of Natural Toxins, Vol. 5. Reptile Venoms and Toxins*, Tu, A. T., Ed., Marcel Dekker, New York, 1991, 147.
21. **Longenecker, G. L.**, Effects of snake venoms on blood platelets, in *Handbook of Natural Toxins, Vol. 5. Reptile Venoms and Toxins*, Tu, A. T., Ed., Marcel Dekker, New York, 1991, 189.
22. **Pirkle, H. and Theodor, I.**, Thrombinlike enzymes from snake venoms, in *Handbook of Natural Toxins, Vol. 5. Reptile Venoms and Toxins*, Tu, A. T., Ed., Marcel Dekker, New York, 1991, 225.
23. **Rosenfeld, G., Nahas, L., and Kelen, E. M. A.**, Coagulant, proteolytic and hemolytic properties of some snake venoms, in *Venomous Animals and Their Venoms, Vol. 1. Venomous Vertebrates*, Bucherl, W., Buckley, E. E., and Deulofeu, V., Eds., Academic Press, Orlando, FL, 1968, 229.
24. **Kisiel, W.**, Effect of snake venoms on factor V, in *Handbook of Natural Toxins, Vol. 5. Reptile Venoms and Toxins*, Tu, A. T., Ed., Marcel Dekker, New York, 1991, 253.
25. **Meier, J. and Stocker, K. F.**, Snake venom protein C activators, in *Handbook of Natural Toxins, Vol. 5. Reptile Venoms and Toxins*, Tu, A. T., Ed., Marcel Dekker, New York, 1991, 265.
26. **Tu, A. T.**, Tissue damaging effects by snake venoms, in *Handbook of Natural Toxins, Vol. 5. Reptile Venoms and Toxins*, Tu, A. T., Ed., Marcel Dekker, New York, 1991, 297.
27. **Slotta, K. H., Gonzalez, J. D., and Roth, S. C.**, The direct and indirect hemolytic factors from animal venoms, in *Animal Toxins*, Russell, F. E. and Saunders, P. R., Eds., Pergamon Press, Oxford, England, 1967, 369.
28. **Chugh, K. S. and Sakhuja, V.**, Renal disease caused by snake venom, in *Handbook of Natural Toxins, Vol. 5. Reptile Venoms and Toxins*, Tu, A. T., Ed., Marcel Dekker, New York, 1991, 471.
29. **Stocker, K.**, Composition of snake venoms, in *Medical Use of Snake Venom Proteins*, Stocker, K. F., Ed., CRC Press, Boca Raton, FL, 1990, 33.
30. **Kockva, E. and Gans, C.**, The structure of the venom gland and secretion of venom in viperid snakes, in *Animal Toxins*, Russell, F. E. and Saunders, P. R., Eds., Pergamon Press, Oxford, England, 1967, 195.
31. **Klauber, L. M.**, Susceptibility, resistance and immunity to rattlesnake venom, in *Rattlesnakes*, Klauber, L. M., Ed., University of California Press, Berkeley, 1972, 883.
32. **Philpot, V. B., Jr. and Stjerholm, R. L.**, Resistance of reptiles to venoms, in *Diseases of Amphibians and Reptiles*, Plenum Press, New York, 1984, 693.
33. **Nicoll, A. A., Douglas, V., and Peck, L.**, On the immunity of rattlesnakes to their venom, *Copeia*, 1933, 211, 1933.
34. **Swanson, P. L.**, Effects of snake venoms on snakes, *Copeia*, 1946, 242, 1946.
35. **Ovadia, M. and Kochva, E.**, Neutralization of Viperidae and Elapidae snake venoms by sera of different animals, *Toxicon*, 15, 541, 1977.
36. **Philpot, V. B. and Smith, R. G.**, Neutralization of pit viper venom by king snake serum, *Proc., Soc. Exp. Biol. Med.*, 74, 521, 1950.
37. **Minton, J. L., Jr. and Minton, M. R.**, *Venomous Reptiles*, Revised ed., Charles Scribner's Sons, New York, 1980.
38. **Werner, R. M. and Vick, J. A.**, Resistance of the opossum *Didelphis virginiana* to envenomation by snakes of the family Crotalidae, *Toxicon*, 15, 29, 1977.
39. **DeWit, C. A. and Westrom, B. R.**, Venom resistance in the hedgehog, *Erinaceus europaeus*: purification and identification of macroglobulin inhibitors as plasma antihemorrhagic factors, *Toxicon*, 25, 315, 1987.
40. **Khole, V.**, Toxicities of snake venoms and their components, in *Handbook of Natural Toxins, Vol. 5. Reptile Venoms and Toxins*, Tu, A. T., Ed., Marcel Dekker, New York, 1991, 405.

Chapter 14

TREATMENT OF SNAKEBITE

I. INTRODUCTION

The history of snakebite treatment is a reflection of the state of knowledge, philosophy, and mythology of humankind through the ages. Perhaps more folklore is associated with snakes and snakebite than any other aspect of medicine. For thousands of years, fear and hatred of snakes, coupled with superstition, religious tradition, and ignorance have prevented a logical approach to the study of snake venoms and snakebite treatment.[1,2]

Handling rattlesnakes was a tradition of some southwestern U.S. native American tribes.[1] Even today, religious cults in the U.S. may demonstrate faith by handling venomous snakes, basing their actions on the biblical statement found in Mark 16:17–18: "And these signs shall follow them that believe; In My name shall they cast out devils; they shall speak with new tongues; They shall take up serpents..." The tenet is that a venomous serpent has no power to harm one who is living righteously.

In the Old Testament is one of the earliest written records of snakebite treatment: "And the Lord said unto Moses: 'Make thee a fiery serpent and set it upon a pole; and it shall come to pass, that every one that is bitten, when he looketh upon it, shall live.' And Moses made a serpent of brass, and put it upon a pole, and it came to pass, that if a serpent had bitten any man, when he beheld the serpent of brass, he lived." — Numbers 21:8–9

Even in modern times, folklore persists. Rural people in underdeveloped countries (where most human and animal snakebite is likely to occur) frequently are poorly educated. Primary modern health care may be minimal or unavailable. People rely on folk cures and native medical practitioners. Many in these environments know of a member of their family or an acquaintance who was injured or died of snakebite. Family stories perpetuate fear and concern. With little factual information and no modern treatments available, it is little wonder that they resort to folk cures.

Although the primary objective of this volume is to discuss venomous bites in animals, the literature of animal envenomation is sparse and mostly anecdotal. Much can be gained from understanding modern approaches to human snakebite management.[1-7]

II. GENERAL MANAGEMENT — HUMAN CROTALID SNAKEBITES

A. FIRST AID

Recommendations for first aid for snakebite have changed materially during the past decade. First aid manuals, Boy Scout handbooks, and general

TABLE 1
First Aid Procedures for Human Crotalid Snakebite

1. Reassure the victim. Anxiety increases cardiovascular activity and aggravates envenomation. Snakebite needn't be fatal.
2. Remove any constricting or potentially obstructing objects (rings, watches, bracelets, and dentures).
3. Immobilize a limb to prevent muscle contraction that will enhance absorption and distribution of venom.
4. Transport the victim to a medical facility for observation and treatment. Avoid muscular activity (walking, moving) by the victim.
5. Watch for signs of severe envenomation (vomition, dyspnea, shock, or paresis). If nausea occurs and vomition seems imminent, place the victim in a horizontal lateral position to allow the easiest egress for the vomitus.
6. Bring in the snake for positive identification if this can be done safely.

TABLE 2
Unacceptable First Aid Procedures for Human Crotalid Snakebite

1. Use of a tourniquet. The application of a tourniquet has caused far more harm than good in the first aid management of crotalid snakebite.
2. Application of ice (cryotherapy).
3. Incision and suction.
4. Consumption of alcohol.
5. Administration of aspirin (potentiates hemorrhage).

recommendations written before 1980 are generally incorrect. Currently recommended steps for first aid management of crotalid snakebite are listed in Table 1. Cryotherapy and the application of a tourniquet should not be performed and medicolegal ramifications may ensue if used. Other time-honored, yet inappropriate first aid procedures are also not recommended (Table 2).

B. PRELIMINARY OBSERVATION, EVALUATION, AND MEDICAL TREATMENT

An initial physical examination should be conducted and blood collected for laboratory analysis. The victim should be observed for signs of envenomation. If there is good reason to believe that a venomous snakebite occurred, but no signs develop, the victim should be monitored for at least 24 h. Venomous snakes may eject no venom (dry bite), in which case no illness will develop other than possible inflammation from fang penetration.

Blood pressure should be monitored frequently. An intravenous catheter should be placed to gain immediate access to the cardiovascular system for treatment of shock or anaphylaxis, and for antivenin administration. Preparation should be made to provide respiratory assistance, including tracheostomy.

Skin testing for allergy to equine serum is controversial. Such testing is not conclusive for all possible reactions that may occur and the test itself may cause subsequent equine serum sensitivity. Even if there is a positive reaction to the skin test, it may not negate the necessity for administration of antivenin in a life-threatening envenomation. Medicolegal ramifications may dictate that skin testing be done, but the inherent problems should be recognized.

C. ANTIVENIN (ANTIVENOM) THERAPY

Antivenin should be administered only in a medical facility with appropriately trained medical personnel having emergency drugs (epinephrine and corticosteroids) and life-support equipment available. The most specific antivenin possible should be obtained once the identity of the snake is ascertained. Although there are no absolute contraindications for the administration of antivenin, caution should be exercised in victims with a known history of asthma, hay fever, or drug sensitivities.[5]

Some world producers of antivenin are listed in Table 3. A more complete listing may be found in References 7 and 8. An *Antivenom Index* is published by the American Association of Zoological Parks and Aquariums in cooperation with the American Association of Poison Control Centers listing the stocks of various antivenins maintained in zoos exhibiting venomous snakes.[8,9] Contact a regional Poison Control Center or the nearest zoo that has a reptile collection to obtain information.

Antivenins are prepared by hyperimmunization of horses. Initially, a nonlethal dose of venom is injected into a horse. Repeated doses of increasing quantity are administered to stimulate the horse's immune system for maximum production of antibodies to the venins within the venom. Only those substances in the venom that are allergenic are of consequence in antivenin production.

Ultimately, serum is collected, purified as much as required, and tested for potency in experimental animals (usually mice). The final product may be a liquid serum, or lyophilized. Although the allergenicity of antivenins has decreased with increasingly sophisticated technology, there is nevertheless a high prevalence of early anaphylactoid response in people to horse serum or other proteins in the product. Many antivenins are also pyrogenic and body temperature should be monitored during administration.

Antivenin is the only specific therapy for snakebite, yet it is not a panacea. The effectiveness of some antivenins for venoms of certain groups of snakes is not precisely known. Specific antivenins are most efficacious, but some polyvalent products have been used effectively. Antivenin should not be administered unless definite signs of envenomation have appeared. The victim should be monitored for development of one or more of the following signs:

1. Hemorrhage distant from the bite site (gingiva, nares, urinary tract, gastrointestinal tract, or at a venipuncture site)
2. Cardiovascular abnormalities (hypotension, bradycardia, or arrhythmias)
3. Evidence of renal dysfunction (oliguria, elevated blood urea nitrogen and/or creatinine, or hemoglobinuria)
4. Nervous system dysfunction (ptosis, dysphagia, slurred speech, dyspnea, paresis, or paralysis)

TABLE 3
World Sources of Antivenin

Address of producer	Product	Venoms used	Used for others
Institut Pasteur d'Algerie Rue Docteur Laveran Alger, Algérie	Antivipérin	*Cerastes cerastes* *Vipera lebetina*	*C. vipera*
Institut Pasteur Place Charles-Nicolle Casablanca, Morocco	Polyvalent	*C. cerastes* *V. lebetina*	*C. vipera*
Al Algousa Share Alvezara Cairo, Egypt	Polyvalent	*Cerastes* spp. *Naja haje*	
South African Institute for Medical Research P.O. Box 1038 Johannesburg 2000 Republic of South Africa	a. Polyvalent	*Bitis* spp. *Dendroaspis* spp. *Hemachatus* spp. *Naja* spp.	*Causus* sp. *D. viridis*
	b. Monovalent	*Echis carinatus*	*Cerastes* spp.
	c. Monovalent	*Dispholidus typus*	
Wyeth International Ltd P.O. Box 8616 Philadelphia, PA 19101	a. Crotalidae	*Crotalus atrox* *C. adamanteus* *C. durissus* *Bothrops atrox*	*Sistrurus* spp. *Agkistrodon* spp. *Calloselasma* spp. *Crotalus* spp. *Lachesis muta*
or Fort Dodge Labs P.O. Box 518 Fort Dodge, IA 50501	b. Monovalent	*Micrurus fulvius*	
Gerencia General de Biologicos y Reactivos SSA Mariano Escobedo 10	Antiviperin	*Bothrops asper* *Crotalus basiliscus*	*Crotalus atrox* *C. durissus* *C. molosus*

Col. Popotla, Mexico, D.F. CP 11400 Mexico			*Agkistrodon bilineatus*
Universidad de Costa Rica Ciudad Universitari Rodrigo facio San José, Costa Rica	a. Anticoral	*Micrurus nigrocinctus*	*M. fulvius* *M. carinicaudus*
	b. Anti-mipartitus	*M. mipartitus*	
	c. Polyvalent serum	*Crotalus durissus* *Lachesis muta* *Bothrops asper*	All Central American species of crotalid snakes
Instituto Butantan Caixa Postal 65 Sao Paulo, SP, Brazil	a. Anti-botropico	*Bothrops* spp., most Brazilian species	
	b. Anti-lauqético	*Lachesis muta*	
	c. Anti-crotalico	*Crotalus durissus*	
	d. Anti-elapidico	*Micrurus* spp.	
	e. Anti-botropica-laquético (a+b)		
	f. Polyvalent (a+c)		
Ministry of Health Department of Laboratories P.O. Box 6115 Jerusalem 91060, Israel	a. Anti-Echis	*Echis coloratus*	
	b. Anti-vipera	*Vipera palestinae*	
Central Research Institute Kasauli 173 205 (HP) India	a. Anti-cobra	*Naja naja*	*Ophiophagus hannah*
	b. Anti-krait	*Bungarus caeruleus*	*Bungarus fasciatus*
	c. Anti-Russell's viper	*Vipera russelli*	*Echis carinatus*
	d. Anti-saw-scaled viper	*Echis carinatus*	*Vipera russelli*
	e. Polyvalent (a+b+c+d)		
Queen Saovabha Memorial Institute The Thai Red Cross Society Rama IV Road Bangkok, Thailand	a. Cobra	*Naja kaouthia*	*Naja naja*
	b. King cobra	*Ophiphagus hannah*	
	c. Banded krait	*Bungarus fasciatus*	*Bungarus caeruleus*
	d. Russell's viper	*Vipera russelli*	
	e. Malayan pit viper	*Calloselasma rhodostoma*	
	f. Green pit viper	*Trimeresurus albolabris*	*Trimeresurus* spp.

TABLE 3 (continued)
World Sources of Antivenin

Address of producer	Product	Venoms used	Used for others
Shanghai Vaccine and	a. Mamushi	*Agkistrodon halys*	
Serum Institute	b. Hundred pace	*Deinagkistrodon acutus*	
1262 Yang An Road (W)	c. Krait	*Bungarus multicinctus*	
Shanghai, China	d. Cobra	*Naja naja*	
Chemo-Sero-Therapeutic	a. Habu antivenom	*Trimeresurus flavoviridis*	
Research Institute	b. Mamushi antivenom	*Agkistrodon halys*	
668 Okubo Shimizu			
Kumamoto 860, Japan			
Pasteur Vaccin	a. Ipser Europe	*Vipera aspis*	*Vipera ursinii*
3 Avenue Pasteur		*V. berus*	
92430 Marnes-La-Coquette		*V. ammodytes*	
France	b. B.E.N.	*Bitis gabonica*	*Bitis nasicornis*
		B. arietans	*Echis coloratus*
		Echis carinatus	Other African *Echis* spp.
		Naja haje	Other African *Naja* spp.
		N. nigricollis	
		N. melanoleuca	
	c. Ipser Afrique	B.E.N. plus	
		Dendroaspis viridis	
		D. jamesoni	
	d. Antirept Pasteur	*Bitis arietans*	
		Echis carinatus	
		Cerastes cerastes	
		Vipera lebetina	
		Naja haje	

Beringwerke AG	a. Europe		
Postfach 11 40		*Vipera aspis*	
3550 Marburg, Germany		*V. ammodytes*	
		V. berus	
		V. lebetina	
	b. Central Africa	*V. xanthina*	
		Bitis gabonica	
		B. arietans	
		B. nasicornis	
		Dendroaspis viridis	
		D. polylepis	
		Hemachatus haemachatus	
		Naja haje	
	c. North and West Africa	*Bitis* spp.	
		Cerastes spp.	
		Vipera lebetina	
		Echis carinatus	
		Naja spp.	
Commonwealth Serum	a. Death adder	*Acanthophis antarcticus*	*Acanthophis pyrrhus*
45 Poplar Road	b. Taipan	*Oxyuranus scutellatus*	
Parkville, Victoria 3052	c. Eastern brown snake	*Pseudonaja textilis*	*Pseudonaja affinis*
Australia			*P. nuchalis*
	d. Brown snake	*Pseudoechis australis*	*P. porphyriacus*
	e. Tiger-sea snake	*Notechis scutatus*	*Austrelaps superba*
		Enhydrina schistosa	*Tropidechis carinatus*
			Pseudoechis spp.

TABLE 3 (continued)
World Sources of Antivenin

Address of producer	Product	Venoms used	Used for others
	f. Polyvalent	*A. antarcticus*	*P. affinis*
		O. scutellatus	*P. nuchalis*
		P. textilis	*A. superba*
		P. australis	*P. porphyriacus*
		N. scutatus	*P. papuanus*

From Chippaux, J.-P., *Handbook of Natural Toxins, Vol. 5. Reptile Venoms and Toxins*, Tu, A. T., Ed., Marcel Dekker, New York, 1991, 529.

5. Massive local swelling (hemorrhage, edema, or vesication)
6. Blood dysfunction (defective blood clotting, or hemolysis)

Once signs of envenomation are evident, intravenous administration of antivenin should commence. Antivenin is usually mixed with physiologic saline or other fluid at approximately 1/500 dilution before initial administration. Antivenin may be administered undiluted if severe envenomation is present or little response is seen to the preferred diluted form.

Response to slow intravenous drip of antivenin should be monitored. Appropriate response should include the cessation of hemorrhage, improvement in blood clotting factors (as determined by laboratory analysis), and clearing of the urine if hemoglobin or myoglobin has been present. Also, blood pressure should stabilize and signs of nervous system dysfunction should improve. If a satisfactory response has not occurred within an hour, more antivenin should be administered, either as a slow bolus or a more rapid drip.

How much antivenin should be administered? The amount of antivenin required is correlated with the dose of venom injected by the snake. Unfortunately, there is no way of measuring the injected dose except by evaluation of the severity of clinical signs of envenomation. With North American crotalid bites, if the physician opts for antivenin therapy, five (5) vials of Antivenin Crotalidae Polyvalent are recommended as the initial dose and 15 vials or more may be required to counteract the effects of envenomation.[6] There is no maximum dose of antivenin. As many as 75 vials have been used to treat a small child bitten twice by an eastern diamondback rattlesnake, *Crotalus adamanteus*.[6] The recommended dose for other antivenins varies from 3 to 15 vials.[1,2,6]

The Mojave green rattlesnake *(Crotalus scutulatus)* produces a highly potent venom. If this snake is identified and envenomation signs develop, the initial administration should be 10 vials of Wyeth Antivenin (Crotalidae) polyvalent rather than 5.[6]

D. REACTIONS TO ANTIVENIN

A victim may have an anaphylactoid reaction to either equine serum or other substances in the antivenin. Modern technology has eliminated many foreign substances from biologic products, but considerable proteinaceous material remains in antivenin. Reactions may immediately follow undiluted intravenous antivenin administration (1 to 20 min) or be delayed slightly (30 to 180 min) when diluted antivenin is administered as a drip.[5]

Initial signs of anaphylaxis include a sensation of heat, restlessness, coughing (mild pulmonary edema), nausea, vomiting, and itching scalp. Untreated anaphylaxis may result in urticaria, generalized itching, fever, tachycardia, and, ultimately, hypotension, bronchospasm, and severe pulmonary edema.[5]

The treatment for anaphylaxis is to stop administration of antivenin until signs can be controlled with epinephrine, given subcutaneously, or, if necessary, intravenously. Corticosteroids and antihistamines have also been given, but antihistamines must be used with caution because some have a hypotensive action.

Serum sickness is a delayed reaction, occurring 5 to 24 d following administration of antivenin. The patient may develop urticaria, fever, and arthralgia, which is usually responsive to antihistamine and corticosteroid therapy. Negative results of skin testing do not preclude later development of serum sickness.[6]

E. MISCELLANEOUS THOUGHTS ON ANTIVENIN THERAPY

Children should receive the same initial dose of antivenin as adults because the amount of venom injected may be the same in children as in adults. In addition, a child's smaller size may allow a higher concentration of venom to attack body tissues. Antivenin should be administered intravenously unless it is impossible to obtain venous access, in which case antivenin may be administered intramuscularly. Injecting antivenin into or near the bite site should be avoided and antivenin should never be injected into a digit.

It is never too late to administer antivenin. Toxins may be in circulation for many days before being excreted by the kidney. However, it should be recognized that once neurotoxins have damaged pre- or post-synaptic nerve impulse transmission, antivenin may fail to reverse these effects. If respiratory paralysis has occurred, forced respiration may be necessary to sustain life.

A special problem arises if a tourniquet has been applied too tightly or left on too long. Immediate release of the tourniquet may flood the circulatory system with toxins, resulting in catastrophic signs of envenomation. To prevent this, intravenous administration of antivenin should be maintained for 3 min before the tourniquet is released. A blood pressure apparatus cuff should be placed above the tourniquet and inflated to above diastolic blood pressure (100 mm mercury). The tourniquet may then be released and the pressure in the cuff slowly decreased while monitoring the patient for worsening signs of envenomation.[5]

III. GENERAL MANAGEMENT OF SNAKEBITE IN ANIMALS

A checklist for managing suspected crotalid envenomation is provided in Table 4.

A. FIRST AID

1. Crotalid Bites

In a few instances a bite may be witnessed, but more often the animal owner will notice alarming signs (either local or systemic) and call a veterinarian. It is seldom possible to positively identify the snake. The animal owner should keep the animal as quiet as possible and seek medical assistance.

TABLE 4
A Checklist for Management of Crotalid Snake Envenomation

A CHECKLIST FOR MANAGEMENT OF CROTALID SNAKE ENVENOMATION

Obtain as much history as possible
- Identity of snake if possible ______________________________
- Size of snake ______________________________
- Time of bite ______________________________

Initial examination
- Local swelling, ______Yes, ______No, How long ______________
- Progressive swelling, ______Yes, ______No, Location ______________
- Pain on palpation, ______Yes, ______No
- Fang marks, ______Yes, ______No, Width between ______________
- Hemorrhage, ______Yes, ______No, Location ______________
- Bullae formation, ______Yes, ______No, Location ______________
- Temperature ______, Heart rate______, Respiration______________
- Blood pressure ________Sys, ______ Dia

Clinical Pathology
- Hematology
- Serum biochemistry
    - Clotting panel
    - Fibrinogen
    - Protein, electrolytes, BUN, creatinine, enzymes
- Urinalysis — hematuria, myoglobinuria, glycosuria

Monitor patient
- Cardiovascular stability — shock, pulse, ECG, blood pressure
- Renal function — urine flow, BUN/creatinine
- Vomiting, diarrhea
- Nerve dysfunction — ptosis, paresis, paralysis, seizures
- Coagulation stability — hemorrhage
- Response to antivenin — usually within an hour

Therapy
- Antivenin (Crotalidae) Polyvalent 3 to 5 vials initially i.v. (not until signs of envenomation appear)
- Broad-spectrum antibiotics
- Analgesics +/–
- Corticosteroids +/–
- Fluids — normal saline or Ringer's
- Supplemental oxygen or assisted respiration if needed (be prepared to perform tracheostomy)
- For emergencies (anaphylaxis)
    - Epinephrine
    - Lidocaine (arrhythmias)
- Restrict activities of the victim
- Local wound treatment if necrosis and sloughing occur
- Tetanus antitoxin/toxoid

2. Elapid or Hydrophid Snakebite

Basic recommendations are the same as for a crotalid bite. However, if the victim cannot be transported to a medical facility within an hour, a tourniquet should be applied proximal to the bite and loosened for 1 min every 10 min. The tourniquet should not be removed until 3 min after antivenin therapy has begun.

B. HOSPITAL CARE

Snakebite may not be suspected by an owner, and, depending on the region and prevalence of snakebite in the area, the veterinarian may not consider snakebite initially. The differential diagnosis may include insect bites or stings, foreign body penetration, puncture wounds, grass awn migration, and contusion.

If venomous snakebite is suspected, a thorough physical examination should be conducted. Blood samples should be collected for hematology and serum chemistry studies. In the case of small companion animals, an intravenous catheter should be placed so that infusion of antivenin may begin immediately if signs of envenomation appear.

Antivenin is the only specific therapy for snake envenomation of animals. Too often, the use of antivenin has been neglected in veterinary medicine. There may be economic constraints (the cost of five vials of antivenin for initial treatment for a crotalid bite is approximately $700.00 in the U.S.), but the rationale for use of antivenin is as valid for an animal as for a human. Antivenins are less expensive in countries other than the U.S., but may be more difficult to obtain.

Antivenin should be administered only if signs of envenomation appear, as described for humans. Skin testing is not usually performed in non-human animals; however, emergency medications and equipment should be on hand to treat anaphylaxis should it develop.

C. TRACHEOSTOMY

Maintaining a patent airway may be difficult in nose and face bites in species that are obligate or semiobligate nasal breathers, as swelling may occlude the nostrils. Preparations should always be made to perform a tracheostomy or place an endotracheal tube if the patient is comatose.

1. Surgery

A 6- to10-cm incision is made on the ventral cervical midline in the middle third of the neck. Preparation of the surgical field, including shaving, cleansing, and administration of a local anesthetic, should be performed if time permits; but in an emergency, forego this and place the tube. The paired sternohyoideus muscles should be separated by blunt dissection (Figure 1). The underlying fascial layer is separated with blunt Mayo scissors or may be incised with a scalpel. No vital structures lie between the skin and the trachea directly on the midline. Avoid scoring the tracheal rings.

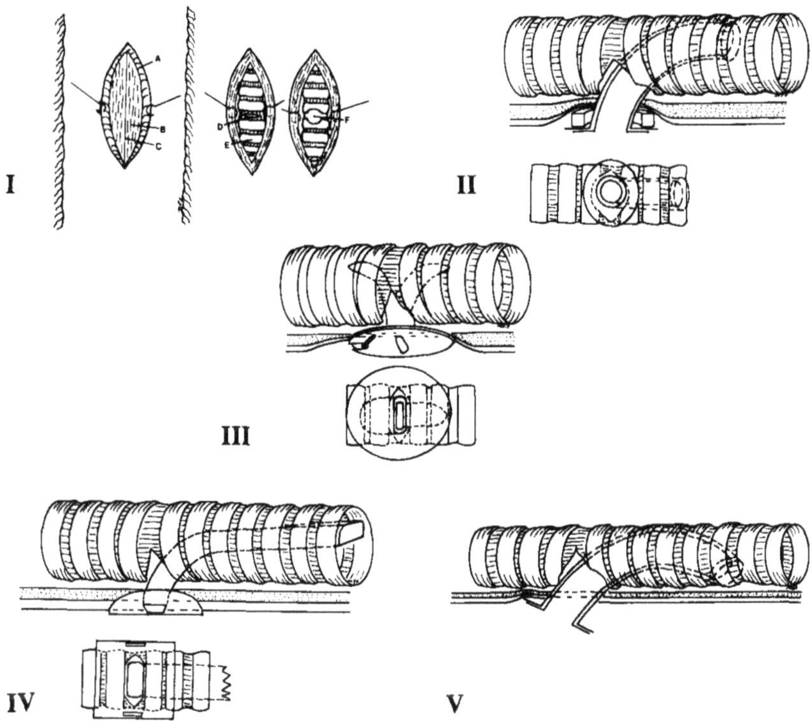

FIGURE 1. Diagrams of: I, surgical approach for a tracheostomy; II, proper placement of a Jackson double-tube tracheostomy tube; III, placement of a Dyson tracheostomy tube; IV, placement of a McKillip tracheostomy tube; and V, improper placement of a Jackson tube, with the tip of the tube against the wall of the trachea. A, retracted skin; B, sternohyoideus muscle; C, midline between the paired muscles; D, incision through the annular ligament; E, tracheal ring; F, removal of semilunar segments from contiguous tracheal rings.

The trachea should be opened by incising the annular ligament between two tracheal rings (Figure 1). If additional space is required for placement of the tracheostomy tube, semilunar segments of contiguous cartilaginous rings may be removed, but avoid incising more than half the width of the ring.

A variety of shapes and sizes of tracheostomy tubes are available and required for the trachea of animals ranging in size from small companion animals to the horse. Many smaller tubes have a semilunar shape. Caution is necessary when placing one of these tubes to ensure that the inner end of the tube does not press against the wall of the trachea. It may be necessary to withdraw the tube slightly and use a spacer to elevate the proximal tube above the skin surface. The tube should have an inner and outer sleeve to facilitate cleaning.

When the tube is fixed in position and the neck bandaged appropriately, a gauze filter should cover the orifice to avoid inhalation of dust or other foreign materials directly into the trachea.

If an appropriately sized tracheostomy tube is not available, an endotracheal tube may be inserted through a tracheostomy incision. If there is a possibility that pulmonary edema will ensue, installation of an endotracheal tube may be necessary so that the cuff can be inflated to aid in providing positive pressure respiratory assistance.

When the need for the tracheostomy is past, the tube should be removed and the wound left to heal by granulation tissue. The wound should be kept covered until the opening into the trachea has closed to avoid inhalation of foreign material.

2. Endotracheal Tube Placement

If the patient is comatose upon arrival at a hospital, assisted respiration via an orally placed endotracheal tube may be crucial to saving the life of the victim. There are three methods of tube placement: (1) blind insertion (the method usually employed in horses); (2) visual insertion while viewing the glottis with a laryngoscope; and (3) placement of a catheter into the trachea and threading the endotracheal tube over the catheter.

The first two methods are standard procedures and veterinarians experienced with companion animals or horses are versed in such placement. The third method has a broad application in domestic and wild animals ranging in size from a mouse to an elephant and should be understood if nontraditional animals are to be given proper medical care in certain types of snake envenomation.

This technique is particularly useful in animals in which the anatomy of the mandibles, tongue, oropharynx, and soft palate make it difficult or impossible to visualize the glottis with a laryngoscope while inserting the tube. The following description is for placement in a llama, *Lama glama*.

Couple two 8-10 French, 50-cm stiff polyethylene catheters together end to end. A long-bladed (45 cm) laryngoscope is optimal, but shorter-bladed scopes may be used if the unit is inserted up to the commissure of the mouth. The head should be extended maximally. It may be desirable to hold the mouth open with gauze loops and gently pull the tongue rostrally. Avoid direct contact of the laryngoscope blade with the epiglottis unless the animal has been anesthetized, is comatose, or the glottis has been sprayed with lidocaine. Otherwise, reflex action may stimulate regurgitation. The polyethylene catheter is inserted into the trachea and the laryngoscope withdrawn. The endotracheal tube is threaded over the catheter and gently inserted between the dental arcades into the trachea.

IV. TREATMENT OF CROTALID BITES IN ANIMALS

Corticosteroids and broad-spectrum antibiotics are frequently administered to cover other possible causes of an inflammatory response.[10,11] Corticosteroids are not effective in countering the direct cytotoxic effects of crotalid snake venom. However, some snake venoms stimulate a histamine response in the victim and corticosteroids may be of value in such instances.[2]

V. MISCELLANEOUS TREATMENTS

Many fanciful or humorous forms of therapy can be discarded without comment, but others have been advocated by medically trained individuals, based on anecdotal experiences.[1,2] Experimental verification of the value of a given therapeutic regimen is lacking for all but a few drugs. That so many drugs and treatment regimens have been advocated is accounted for by the fact that the severity of envenomation varies tremendously from case to case, and recovery may occur, without treatment, in 3 to 5 d. In such cases, any form of therapy may be given credit for the recovery.[2]

A. REGIMENS OF QUESTIONABLE VALUE

The administration of calcium gluconate is recommended for black widow spider envenomation, but there is no evidence that it is effective in snakebite. Heparin has been administered to prevent disseminated intravascular coagulation, but the reverse, with decreased clotting time, is just as probable a result of envenomation, making the administration of heparin hazardous. Dimethyl sulfoxide (DMSO) has been advocated for both local and intravenous therapy to reduce the swelling associated with snakebite.[12] Swelling in snakebite is not an inflammatory response, but a cytotoxic effect on the endothelium of vessels, and is not likely to be responsive to DMSO. Other medications of questionable value include ascorbic acid and ethylenediaminetetraacetic acid (EDTA, a substance often used to prevent coagulation in blood samples). EDTA inactivates venom enzymes, and was thought to reduce local tissue damage if injected at the bite site within a few minutes of the bite. Although advocated for a time, the use of EDTA has been abandoned.[2]

Potassium permanganate is known to neutralize snake venom *in vitro,* but there is no evidence that such a reaction occurs *in vivo.* Used in a dilute solution, permanganate is an antiseptic; but when injected into tissue at the bite site or if crystals are rubbed into fang wounds, the caustic effect of permanganate may destroy tissue.[2] Antihistamines have been used in treatment of human snakebite, particularly to prevent the development of anaphylactoid responses, but caution should be exercised, as the hypotensive action of some antihistamines may act in an additive manner with the action of venoms.[13,14] Low-voltage electrical stimulation has been advocated recently, but experimental studies have produced negative results.[15-17]

1. Contraindicated Forms of Therapy

The use of a tourniquet or incision and suction cannot be recommended for either first aid or hospital therapy in cases of crotalid envenomation.[18] The development of local infection and septicemia or gangrene too often result from these practices. Cryotherapy (the application of ice or ethyl chloride spray to the bite site)[19] is also not acceptable.[1] Although promulgated and advertised in outdoor sporting magazines, there is no experimental evidence to show that this form of therapy has any value and numerous lawsuits have arisen because of tissue destruction and disability resulting from cryotherapy.[1]

The application of cold compresses or cold hydrotherapy is frequently advocated in the treatment of crotalid snakebite in horses and livestock. Again, the results of experimental studies indicate that this practice probably has no value and may be contraindicated.[1]

VI. PREVENTION OF SNAKEBITE

Logically, avoiding encounters with venomous snakes is the best method of prevention. Humans can do much to minimize risk, but some risk is unavoidable if persons venture into the wilderness. As stated elsewhere in this volume, there may be a risk even in residential areas that have been recently developed in venomous snake habitat.

Zoo veterinarians may be called upon to eliminate the danger of snake envenomation to zoo keepers by removal of the venom gland or by ligation of the venom duct. Although in the past this has been attempted at one or more zoos, the author is unaware of any zoo that now employs devenomation on a routine basis. Devenomation is surgically possible, but there are many ramifications that should be considered before embarking on a devenomation protocol. Snake venom has a digestive function, and if no longer injected into a rodent, digestion may be less than optimal. The stress of the surgery must also be considered.

Devenomation should never be used as an alternative to proper venomous snake handling and management. Neither should such a procedure be performed on a privately owned snake. The legal ramifications are manifold. The author once ligated the venom ducts of a rattlesnake that was used to demonstrate to students how to handle venomous snakes for medical purposes. Immediately after the operation, the snake would strike at rodents offered for food, and wait for the mouse to die. After repeated unproductive strikes, the snake finally grasped the rodent and fed normally. Generally, the snake was fed dead prey, but periodically a live mouse was offered. After several months, an attentive technician noticed that the live-food mouse had been killed by the snake strike. Recannulation of the venom duct had occurred. No liberties had been taken with this snake; but if it had been a privately owned snake and a bite had occurred, it is easy to see the liability involved.

Removal of the fangs is useless as a preventive measure because replacement fangs appear within a short time.

A. SURGICAL DEVENOMATION

1. Ligation of the Venom Duct

Devenomation may be an acceptable procedure for certain types of experimental investigation on venomous snakes. To ligate the venom duct, the snake should be under deep general anesthesia and the head secured in a dorsolateral position to expose the upper lip. The duct may be exposed via an incision on the mucosal side of the upper lip, approximately ventral and slightly rostral to the orbit. Alternatively, the incision may be made in the skin, 5 mm above the lip in the same location. The duct is isolated by blunt dissection. The area should be thoroughly explored to ascertain the presence of any auxiliary ducts. Double ligate the duct and remove a segment 1 cm long. Incision closure is routine.

The area of the gland swells post-operatively until back pressure inhibits glandular secretion. Always test the effectiveness of the surgery by offering the snake live prey.

2. Removal of the Gland

The removal of the entire gland is much more traumatic for the snake than ligation of the venom duct. The gland lies on the caudolateral side of the head, dorsal to the commissure of the mouth and caudal to the orbit. A curvilinear incision is made over the gland, rostrally sufficient to allow identification of the duct. The gland lies beneath a sheet of muscle that contracts to express the venom from the gland during a strike.

Extirpate the gland by blunt dissection rostrally to the duct, which is then double ligated. The pocket vacated by the gland should be carefully obliterated by placement of absorbable sutures. The skin is sutured with a pyrogallic acid or other absorbable suture material, to avoid the necessity of removal at a later time.

REFERENCES

1. **Russell, F. E.,** *Snake Venom Poisoning,* Scholium International, Great Neck, NY, 1983.
2. **Minton, S. A., Jr. and Rutherford Minton, M.,** *Venomous Reptiles,* Revised ed., Charles Scribner's Sons, New York, 1969.
3. **Reid, H. A. and Theakson, R. D. G.,** The management of snakebite, *Bull. World Health Org.,* 61, 885, 1983.
4. **Russell, F. E.,** Snake venom poisoning, *Vet. Human Toxicol.,* 33, 584, 1991.
5. **Campbell, J. A. and Lamar, W. W.,** *The Venomous Reptiles of Latin America,* Comstock Publishing, Ithaca, NY, 1989.
6. **Wingert, W. A.,** Management of crotalid envenomations, in *Handbook of Natural Toxins, Vol. 5. Reptile Venoms and Toxins,* Tu, A. T., Ed., Marcel Dekker, New York, 1991, 611.
7. **Chippaux, J.-P.,** Production and Use of snake antivenin, in *Handbook of Natural Toxins, Vol. 5. Reptile Venoms and Toxins,* Tu, A. T., Ed., Marcel Dekker, New York, 1991, 529.
8. **Belcher, D.,** Antivenom Index, American Association of Parks and Aquariums and American Association of Poison Control Centers, Wheeling, WV, 1989.
9. **Tong, T.,** *Antivenom Index,* American Association Zoological Parks and Aquariums in cooperation with American Association of Poison Control Centers, Oglebay Park, Wheeling, WV, 1989.
10. **Knapp, W. A. and Flowers, H. H.,** Treatment of poisonous snake bite in the dog with cortisone acetate, *Vet. Med. Small Animal Clin.,* 51, 475, 1956.

11. **Cunningham, E. R., Stabback, M. S., et al.,** Snakebite: role of corticosteroids as immediate therapy in an animal model. *Am. Surg.*, 45, 757, 1979.
12. **Lee, J.,** Dimethyl sulfoxide (DMSO) in experimental treatment of snakebite in dogs, *Vet. Med. Small Animal Clin.*, 76, 404, 1970.
13. **Maier, H. K.,** Benadryl hydrochloride in treatment of snakebite in dogs, *Vet. Med.*, 46, 463, 1951.
14. **Parrish, H. M., Scatterday, J. E., and Pollard, C. B.,** The use of antihistamine (Phenergan) in experimental snake venom poisoning in dogs, *J. Am. Vet. Med. Assoc.*, 129, 522, 1956.
15. **Guderian, R. H., Mackenzie, C. D., and Williams, J. F.,** High voltage shock treatment for snakebite, *Lancet ii*, 229, 1986.
16. **Johnson, E. K., Kardong, K. V., and Mackessy, S. P.,** Electric shocks are ineffective in treatment of lethal effects of rattlesnake envenomation in mice, *Toxicon*, 25, 1347, 1987.
17. **Schmutzhard, E.,** Electric shocks for snakebite, *Lancet ii*, 578, 1986.
18. **Parrish, H. M.,** Early excision and suction of snakebite wound in dogs, *N. Carolina Med. J.*, 16, 93, 1957.
19. **Stahnke, H. L.,** *The Treatment of Venomous Bites and Stings*, Revised ed., Arizona State University, Tempe, 1966.

Chapter 15

SNAKEBITE IN THE U.S.

I. VENOMOUS SNAKES

Approximately 120 species of snakes inhabit the U.S., with only 21 species being venomous.[1] Many venomous species are small, produce low potency venom, or have such an insular or restricted range that animal bite exposure is unlikely. The more important species involved in animal envenomation are listed in Table 1. Distribution of these species is illustrated in figures accompanying a description of the snake. Snakes are not spread in a uniform distribution pattern throughout their range. Suitable habitat, availability of adequate prey, and freedom from encroachment by human activities influence where snakes are found.

Two species of coral snakes *(Elapidae)* are native to the U.S. (Table 1, Figure 1), but they pose little risk to animals because of their small size and secretive behavior. The major dangerous species are pit vipers *(Crotalidae)*, Table 2.

A. DANGEROUS SPECIES

The eastern diamondback rattlesnake *(Crotalus adamanteus)* is the largest crotalid in the U.S., with some specimens reaching a length of 213 cm (7 ft), (Figures 2 and 3).[2] Diamondback venom is not highly toxic (1.54 to 2.4 mg/kg, mouse i.v.); but because of the snake's aggressive behavior and large potential injection dose (up to 683 mg dry weight), a bite in a small animal may be fatal. Most of the cases of severe envenomation of animals within its range are caused by the eastern diamondback.

The western diamondback rattlesnake *(Crotalus atrox)* is only slightly smaller than its eastern cousin (Figures 3 and 4). It also has a pugnacious disposition and produces approximately the same amount of venom with a similar toxicity.[2]

The Mojave (green) rattlesnake, *Crotalus scutulatus* (Figure 5), is a small snake [maximum length approximately 123 cm (4 ft)] that inhabits deserts and mountains from southern California and Arizona into central Mexico (Figure 6). Venom yield varies from 8 to 139 mg dry weight, but the toxicity of Mojave rattlesnake venom (0.14 to 0.21 mg/kg) makes this species the most dangerous crotalid in the U.S.[2]

Together, the nine subspecies of *Crotalus viridis* inhabit most of the western U.S. (Table 1 and Figures 7 and 8). Subspecies vary in maximum length from 70 cm (2 ft) to 175 cm (5.75 ft). Venom yield in larger snakes may reach 200 mg dry weight, with a toxicity of 1.0 to 3.0 mg/kg. Large snakes may produce moderate to severe envenomation in animals.

The timber rattlesnake, *Crotalus horridus horridus* (Figure 9), and the canebrake rattlesnake, *C. horridus atricaudatus*, are both large snakes [maximum

TABLE 1
Venomous Snakes — United States and Canada

Phylum — *Chordata*
Class — *Reptilia*
Order — *Squamata*
Suborder — *Ophidia (Serpentes)* — snakes
  Family — *Elapidae*
    Micruroides euryxanthus — Sonoran (Arizona) coral snake
    Micrurus fulvius fulvius — eastern coral snake
    Micrurus fulvius tenere — Texas coral snake
  Family — *Crotalidae*
    Agkistrodon contortrix — American copperhead
    Agkistrodon piscivorus — water moccasin (cottonmouth)
    Crotalus spp. — rattlesnakes (RS) — 32 species, 70 subspecies in New World[6]
    C. adamanteus — eastern diamondback RS
    C. atrox — western diamondback RS
    C. cerastes — sidewinder RS
    C. horridus horridus — timber RS
    C. horridus atricaudatus — canebrake RS
    C. scutulatus — Mojave (green) RS
    C. viridis helleri — southern Pacific RS
    C. viridis lutosus — Great Basin RS
    C. viridis oreganus — northern Pacific RS
    C. viridis viridis — prairie RS
    Sistrurus catenatus — massasauga RS
    Sistrurus miliarius — pygmy RS

length 190 cm (6.25 ft)], inhabiting the central and southeastern U.S. (Figure 7). Toxicity varies from 1.64 to 3.09 mg/kg.[2]

Water (cottonmouth) moccasins, *Agkistrodon piscivorus*, (Figures 10 and 11) may cause moderate to severe animal envenomation, but the copperhead *A. contortrix* (Figures 11 and 12) rarely causes serious animal bites. Neither does the pygmy rattlesnake (Figure 13) or massasauga *Sistrurus* spp.

II. CONDITIONS OF ENVENOMATION

It is estimated that 45,000 people are bitten annually by snakes in the U.S.; of these, approximately 7000 are treated by a physician for envenomation.[1] Statistics for the prevalence of snakebite in animals are unavailable, but it is reasonable to assume that snakebite in animals is more common than in humans. Most incidents of animal snakebite go undetected and untreated.

Snakebite occurs in every state except Alaska and Hawaii. Venomous snakes may be active year round in the southern tier of states; but in the northern temperate regions, activity is restricted to late spring, summer, and early fall. Inasmuch as snakes are ectothermic, feeding and resting patterns are determined by ambient temperature. In desert regions, snakes tend to be nocturnal to avoid lethal daytime temperatures. The prairie rattlesnake, *Crotalus viridis viridis,* inhabits regions from southern Saskatchewan in Canada to northern Chihuahua in Mexico. Its activity patterns vary with the region.

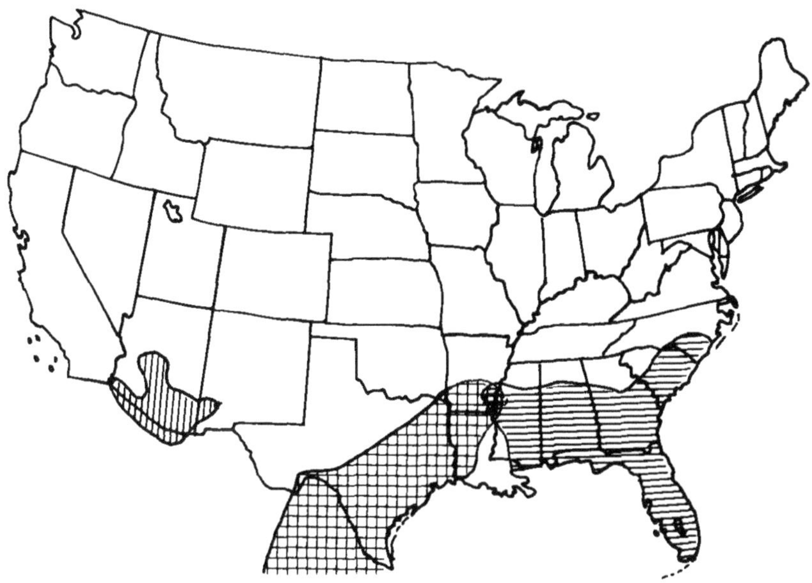

FIGURE 1. Distribution of coral snakes in the U.S. Sonoran coral snake, *Micruroides euryxanthus*. Texas coral snake, *Micrurus fulvius tenere*. Eastern coral snake, *Micrurus fulvius fulvius*. (After Russell, F. E., *Snake Venom Poisoning*, Scholium International, Great Neck, NY, 1983. With permission.)

TABLE 2
Relative Toxicity Data on Snake Venoms of the United States

Common and scientific name	Length (cm)	Approximate dry venom yield (mg)	i.v. LD_{50} mice (mg/kg)
Eastern diamondback rattlesnake *Crotalus adamanteus*	81–165	370–700	1.2–2.4
Western diamondback rattlesnake *Crotalus atrox*	76–193	190–423	1.2–3.5
Mojave rattlesnake *Crotalus scutulatus*	56–102	50–90	0.21
Timber rattlesnake *Crotalus horridus horridus*	81–197	75–150	2.6
Southern Pacific Rattlesnake *Crotalus viridis helleri*	81–122	75–150	1.29
Water moccasin *Agkistrodon piscivorus*	76–127	90–145	4.0
Copperhead *Agkistrodon contortrix*	61–91	40–70	10.9
Eastern coral snake *Micrurus fulvius fulvius*	58–81	2–6	0.38

Modified from Tu, A. T., *Rattlesnake Venoms*, Tu, A. T., Ed., Marcel Dekker, New York, 1982.

FIGURE 2. Eastern diamondback rattlesnake, *Crotalus adamanteus*.

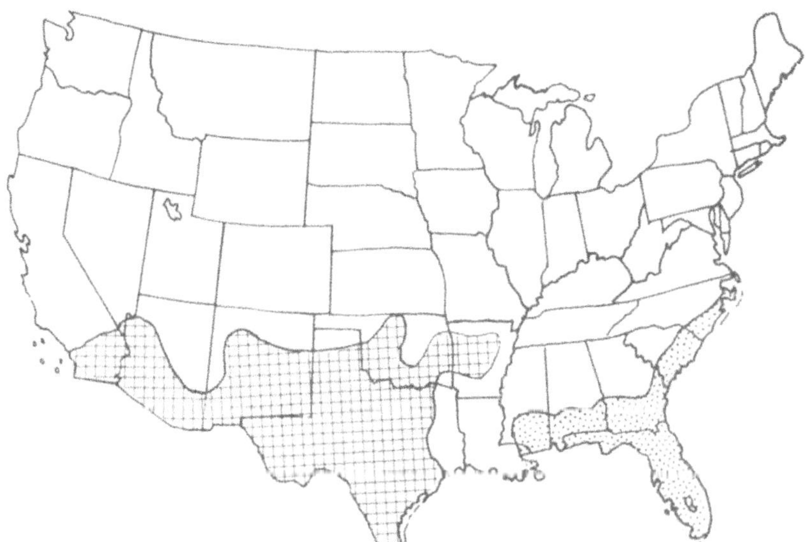

FIGURE 3. Distribution in the U.S. ▒ Eastern diamondback rattlesnake, *Crotalus adamanteus*. ▦ Western diamondback rattlesnake, *Crotalus atrox*. (From Russell, F. E., *Snake Venom Poisoning*, Scholium International, Great Neck, NY, 1983. With permission.)

FIGURE 4. Western diamondback rattlesnake, *Crotalus atrox*. Photo courtesy R. Cogan, Phoenix, Arizona Zoo.

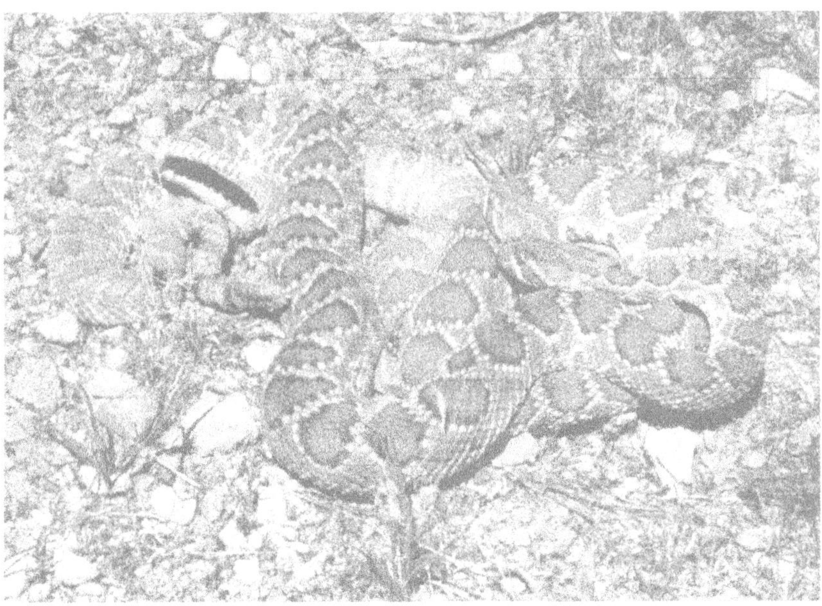

FIGURE 5. Mojave rattlesnake, *C. scutulatus*.

FIGURE 6. Distribution of Mojave rattlesnake, *Crotalus scutulatus*, in North America. Adapted from Campbell, J. A. and Lamar, W. W., *The Venomous Reptiles of Latin America*, Comstock Publishing, Ithaca, NY, 1989, and Russell.[6]

Dogs may be bitten when coursing through fields and woods, especially in tall vegetation. If an inquisitive dog sees a snake, it may move within striking distance to investigate. Young, inexperienced dogs are most likely to approach too closely. Rattlesnakes may not rattle before striking. Yard-confined pets have been bitten by snakes returning to newly established residential areas that were previously snake habitat.

Aggressive dogs may attack a snake and be bitten, as it is not possible for a dog to jump away from a rattlesnake strike. Even the legendary mongoose is incapable of successfully attacking a crotalid. A mongoose may kill a cobra because of the cobra's striking behavior, see page 115, Chapter 13, but a crotalid strike is much too fast to dodge.

It is estimated that several hundred horses are bitten each year.[3,4] Horses may be bitten on the head or neck while grazing in tall grass. Limb bites occur

FIGURE 7. Northern Pacific rattlesnake, *Crotalus viridis oreganus.*

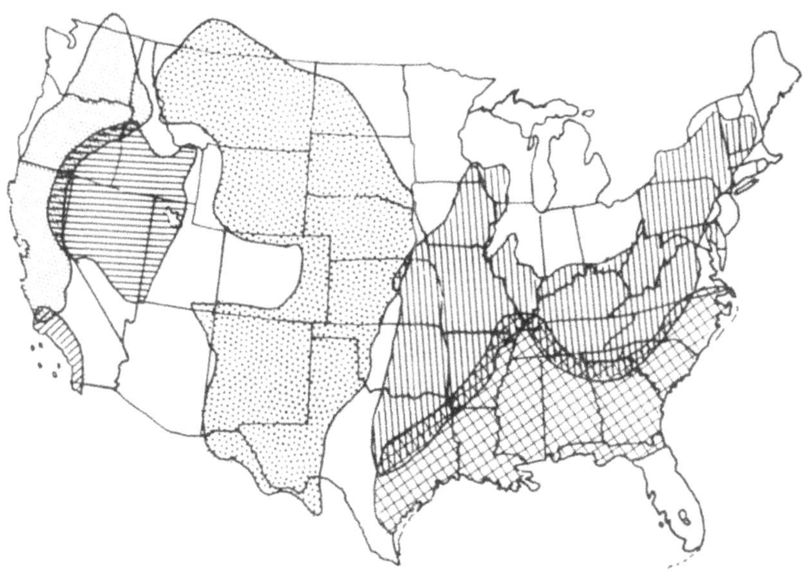

FIGURE 8. Distribution of some rattlesnakes in the U.S. Northern Pacific, *Crotalus viridis oreganus.* Southern Pacific, *Crotalus viridis helleri.* Great Basin, *Crotalus viridis lotosus.* Prairie, *Crotalus viridis viridis.* Timber, *Crotalus horridus horridus.* Canebrake, *Crotalus horridus atricaudatus.* (From Russell, F. E., *Snake Venom Poisoning*, Scholium International, Great Neck, NY, 1983. With permission.)

FIGURE 9. Timber rattlesnake, *Crotalus horridus horridus*.

FIGURE 10. Water (cottonmouth) moccasin, *Agkistrodon piscivorus*.

FIGURE 11. Distribution of ▥ Copperhead, *Agkistrodon contortrix*. ▒ Water moccasin. *Agkistrodon piscivorus*. (From Russell, F. E., *Snake Venom Poisoning*, Scholium International, Great Neck, NY, 1983. With permission.)

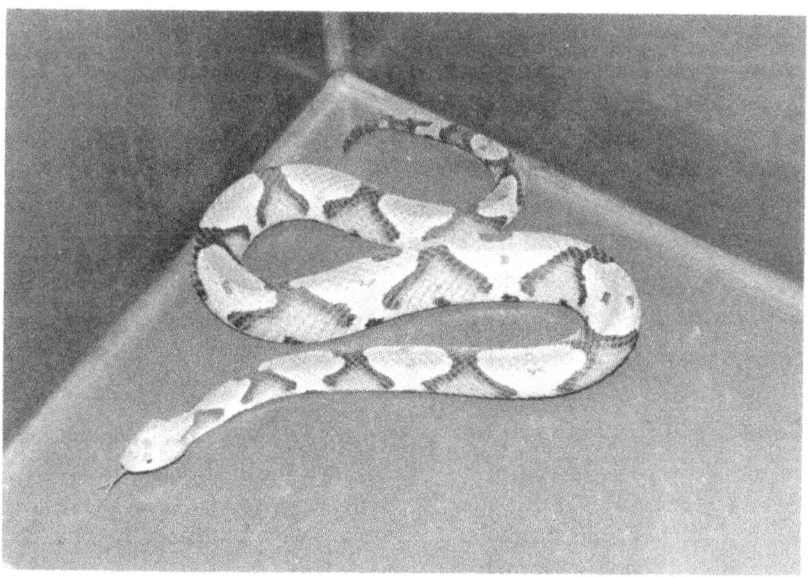

FIGURE 12. Copperhead, *Agkistrodon contortrix*. (Photo by the author, courtesy Steinhart Aquarium, San Francisco, CA.)

FIGURE 13. Pygmy rattlesnake, *Sistrurus* spp.

while walking in tall vegetation or along trails in the wilderness. Inquisitive foals and yearlings frequently experience bites on the nose. Occasionally, a horse may accidentally step on a resting crotalid, but given any kind of warning, the snake tends to move away.

Llamas and alpacas are naturally inquisitive and even adults may approach strange animals in their environment. Venomous snakes are essentially nonexistent in the native habitats of South American camelids; thus, there has been no evolutionary selective pressure to avoid snakes. Nose and face bites are the most common.

Crotalid snakes and non-prey sympatric wild animal species generally coexist without harm, but accidental encounters resulting in bites must surely occur. Captive wild animals have been bitten when snakes entered their enclosures.

III. VENOM

Crotalid venoms tend to be more complex than those of elapid snakes.[2] Rattlesnake venoms contain a high percentage of non-neurotoxic proteins (90% of the dry weight of rattlesnake venom is protein),[5] with proteolytic enzymes, hemorrhagic toxins, and myotoxins predominating. Neurotoxins predominate in coral snake venom.

Two new neurotoxins (crotoxin and Mojave toxin) have been isolated from the neotropical rattlesnake, *Crotalus durissus,* and the Mojave rattlesnake. The venoms of these snakes produce syndromes (described later) differing from the syndrome of typical crotalid envenomation. It is of interest to note that the concentration of Mojave toxin isolated from venom of the Mojave rattlesnake varies markedly from north to south within the snake's distributional range. Snakes in the southern aspects of the range (Mexico) have a much more potent venom.[5]

IV. SIGNS OF ENVENOMATION

A. CROTALID SNAKES

All pit vipers are venomous, but smaller species or those that produce less potent venom may cause minimal clinical signs. Envenomation may be classified as *mild,* with swelling, pain, and erythema occurring only at the bite site within an hour of the bite; *moderate,* if swelling progresses beyond the bite site and there is pain, local hemorrhage, and, possibly, subsequent tissue necrosis; or *severe,* if systemic manifestations develop.

All degrees of envenomation occur from crotalid bites, depending on the injected dose of venom (size of snake, single or multiple bites, and aggressiveness of snake), potency of the venom, type of bite inflicted (single or double fang, scratch, or subcutaneous or deep injection), and the species of animal bitten. The larger the snake and the smaller the victim, the more the likelihood of severe envenomation.

1. Human Crotalid Envenomation[1,6,7]

Usually, one or two fang marks are evident, with a distance between fang marks of 0.5 to 4.0 cm. Swelling may enhance the distance and subsequent estimation of the size of the snake. Most victims experience a burning pain sensation immediately following the bite, followed within a few minutes by swelling (pitting edema) that progresses to involve an entire limb within 6 to 8 h.

Discoloration from subcutaneous hemorrhage appears initially at the bite site, and may progress proximally up the limb. If severe envenomation is untreated, or if treatment is delayed, marked edema with bullae formation or hemorrhagic blebs will develop in 6 to 36 h.

Neurologic signs are minimal in most crotalid bites; but if paresthesia of the scalp, face, and lips or a metallic taste in the mouth occur, severe envenomation is indicated.[1] Other signs of envenomation include weakness, faintness, nausea, and vomiting. Crotalid venom contains a hemotoxin that damages endothelium, allowing hemorrhage to occur both externally and internally at any body site (epistaxis, melena, hematuria, cutaneous petechiae, and ecchymoses).[1] The ultimate effects of severe, untreated envenomation are hypovolemic shock, pulmonary edema, and renal failure.

Human envenomation by the Mojave (green) rattlesnake produces a different syndrome. Little or no immediate pain or swelling may be associated with the bite; but within a few hours, signs of neural dysfunction appear, including diplopia, hoarseness, inability to swallow, and dyspnea associated with progressive respiratory paralysis.[1] Russell[6] felt that neural dysfunction may be a reflection of cerebral hypoxemia associated with severe anemia. However, more recently, a neurotoxin (Mojave toxin) has been identified in Mojave rattlesnake venom.[5] Terminal convulsions may result from the effects of Mojave toxin or simply reflect agonal struggling.[6] This snake accounts for the majority of fatal human snakebites in California. The syndrome associated with this snakebite has not been reported in animals, but probably occurs in dogs under the same circumstances as for human bites.

Bites of humans from other pit vipers (water moccasin, copperhead, pygmy rattlesnake, and massasauga) usually fit into the mild to moderate crotalid envenomation category.

2. Dog Crotalid Envenomation[9-14]

Snakebite of a dog is seldom witnessed; therefore, signs have usually progressed to those of moderate envenomation before the animal is seen by a veterinarian. Schaer[9] described the collective syndrome in a study of a series of eastern diamondback rattlesnake bites in dogs. Dogs were usually depressed and tachypneic upon presentation. Blood was frequently oozing from fang marks on the face, limbs, or body. Significant edema of the head (Figure 14) or limbs and hemorrhage were consistent findings (epistaxis, gingival and labial petechia and ecchymoses, and subcutaneous petechia on the ventral abdomen). Additional signs of moderate envenomation included sinus tachycardia, laryngeal edema, and chemosis.[10,11]

Systemic signs of severe envenomation also included shock, ventricular premature contractions, anuria, hemolysis, ventricular fibrillation, hemorrhagic diathesis (prolonged clotting time and disseminated intravascular coagulation), periorbital cellulitis, panophthalmitis, fibrinolysis, tissue necrosis and sloughing, (Figure 15), and, ultimately, death. All dogs that died had hemorrhagic diathesis or hemolysis.

Mansfield[11] reported similar signs in a series of snake envenomation in dogs, but also observed vomition, diarrhea, hypotension, anorexia, excessive salivation, tremors, coma, pulmonary edema, oliguria or anuria, paralysis, and convulsions.

3. Horse Crotalid Envenomation

In horses bitten on the nose, head, neck, or limb, pitting edema occurs at the bite site and progresses to include the entire head or limb. The author's experience has been with bites from the northern Pacific rattlesnake, *Crotalus viridis oreganus*, which is a relatively small snake that usually produces only mild to moderate envenomation.

In bites of the nose or head, edematous swelling of the nose (Figures 16 and 17) and nasal mucosa may be accompanied by blood-tinged fluid oozing from one or both nostrils.[3,4] The eyelids swell (chemosis) and excessive lacrimal secretions may be tinged with blood. The entire head, including the lips and ears, may become edematous. The pitting edema is not hot to the touch.

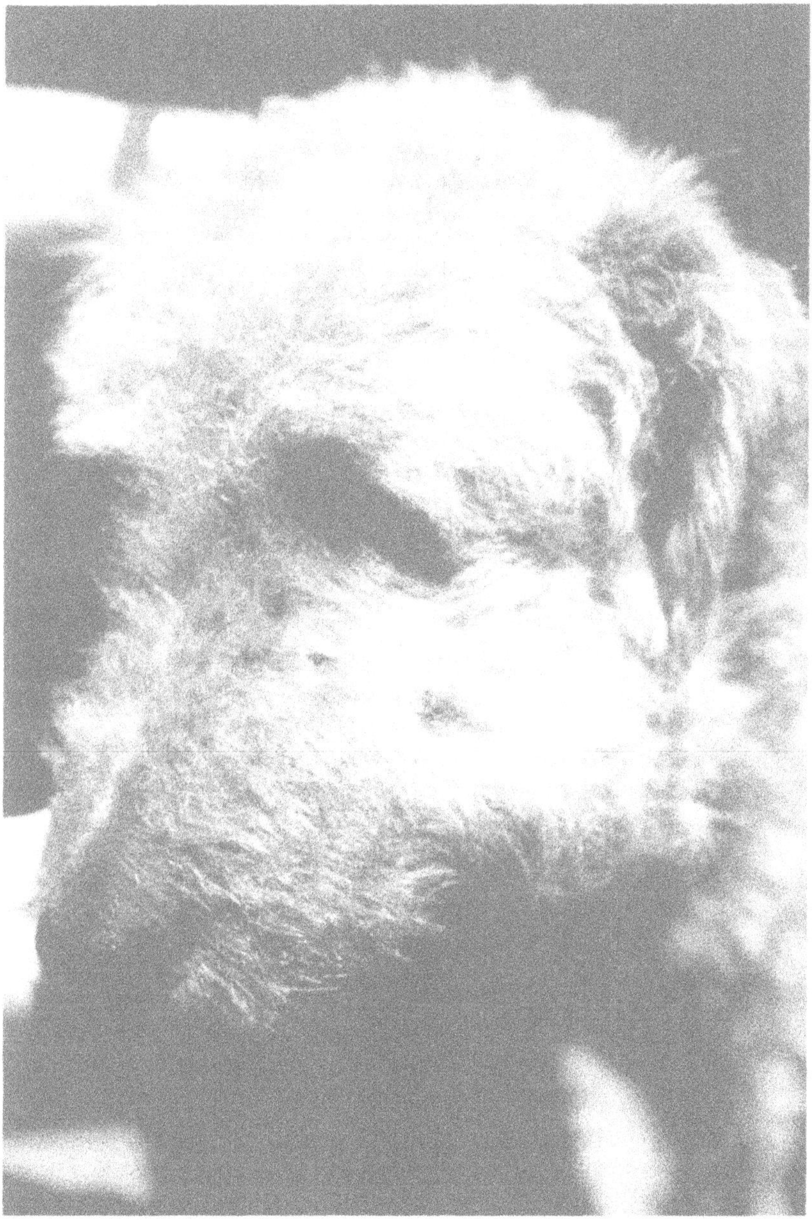

FIGURE 14. Rattlesnake bite on the head of a dog.

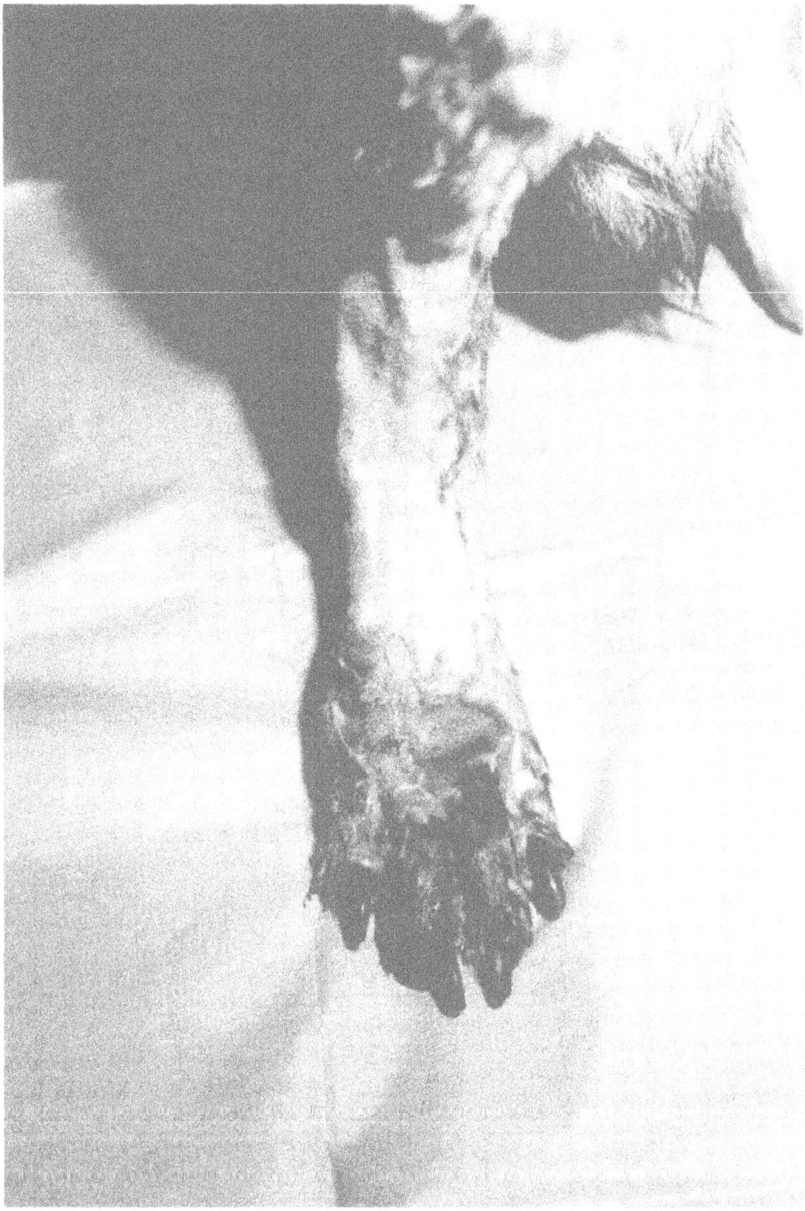

FIGURE 15. Necrosis and sloughing caused by a rattlesnake bite in a dog.

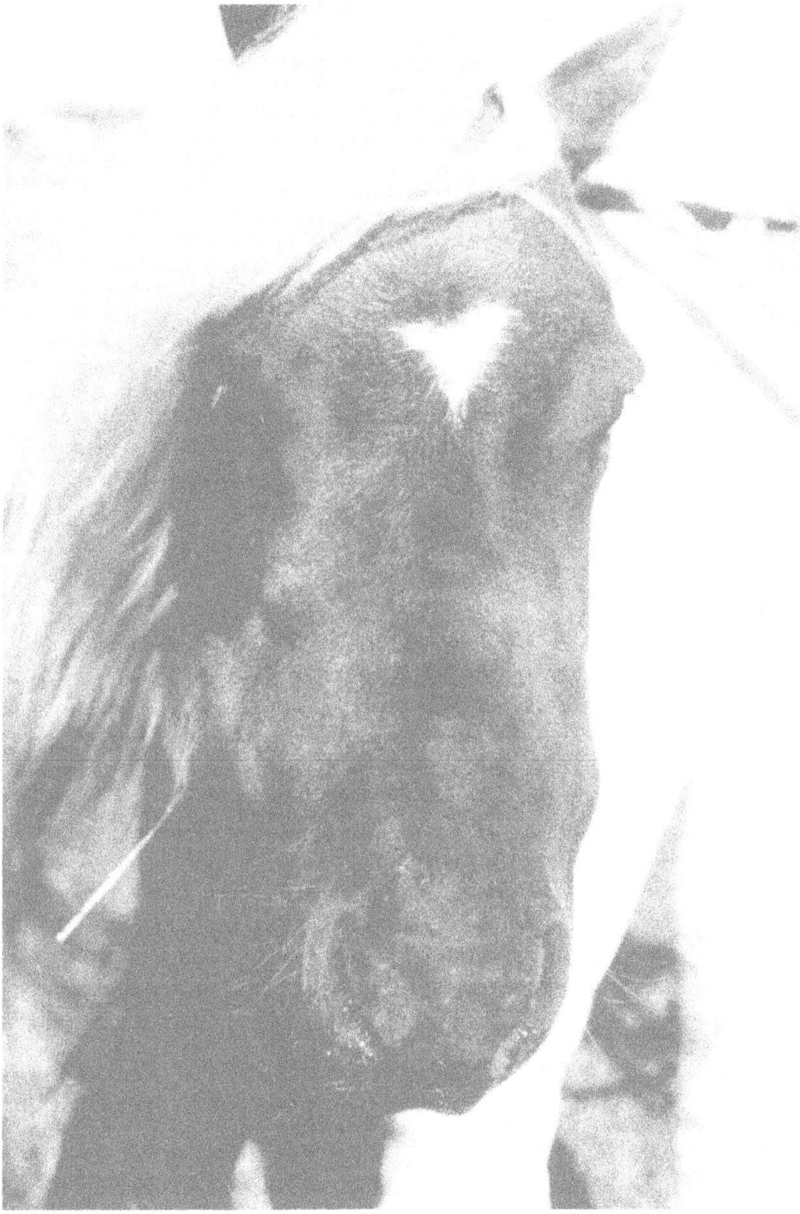

FIGURE 16. Swollen muzzle caused by rattlesnake bite in a pony.

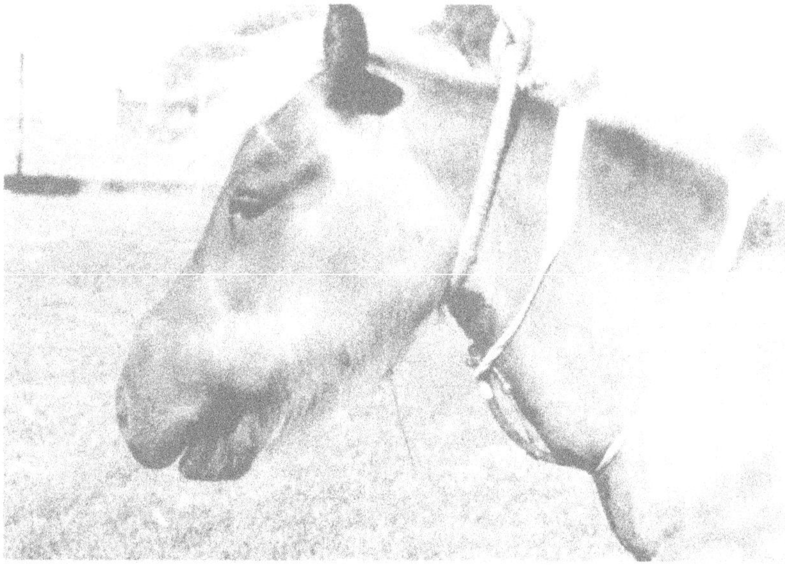

FIGURE 17. Swollen muzzle on a pony following a bite from a northern Pacific rattlesnake, *Crotalus viridis oreganus;* tracheostomy tube in place.

Limb bites are usually unilateral, below the carpus or tarsus. Swelling progresses up the limb to include contiguous areas of the trunk (Figures 18 and 19). Tissue necrosis has been rare in the author's experience, but it has been reported in areas of the country where diamondback rattlesnakes are found.

Dyspnea is a vital sign that must be monitored carefully. The horse is an obligate nasal breather because of an elongated soft palate. Swelling of the nostrils or nasal mucosa occludes the air passageways and may cause suffocation if not observed and treated. Pharyngeal swelling from a neck bite may obstruct air flow at that level. Death from snakebite is rare in horses unless there is hemorrhage into a vital organ or if head bites go untreated and the horse suffocates.

4. Llama Crotalid Envenomation

Signs of snakebite in llamas are similar to the syndrome observed in horses (Figures 20 and 21).[15] Llamas are also obligate nasal breathers and bites on the nose may be fatal within 2 h if a tracheostomy tube is not placed. No tissue necrosis or systemic signs have been observed with envenomation by northern Pacific rattlesnakes, but bites from large diamondback rattlesnakes may cause necrosis if envenomation is severe.

5. Sheep and Other Livestock

Clinical signs are usually restricted to localized swelling of the head (Figures 22 and 23) or limbs.[16-21]

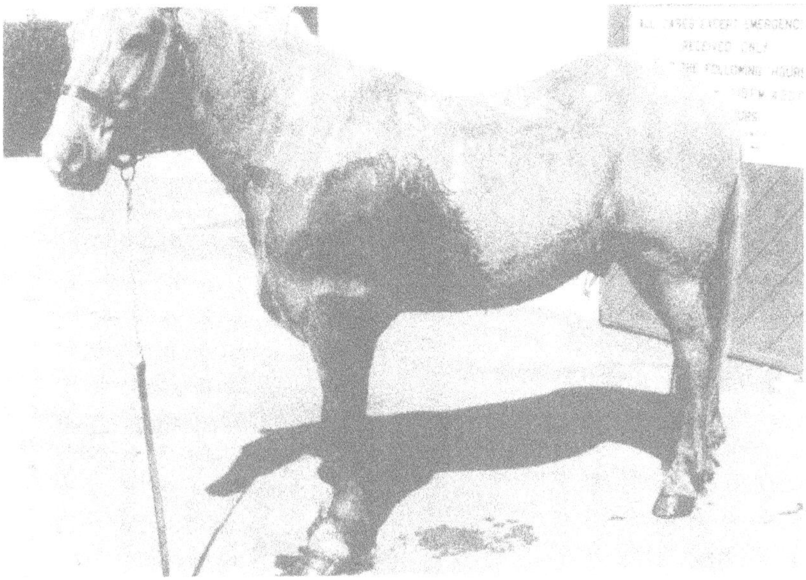

FIGURE 18. Rattlesnake bite of the left carpus.

6. Wild Animals

Accidental envenomation of wild animals other than prey species must surely occur, but diagnosis is rarely made.[22] Envenomation by the prairie rattlesnake occurred in eight captive Rocky Mountain elk, *Cervus elaphus nelsoni*, maintained in a herd in an enclosure in Colorado.[23]

Clinical signs included painful swelling (restricted to the face, muzzle, and submandibular space), inspiratory dyspnea, epistaxis, frothy, blood-tinged nasal discharge, epiphora, anorexia, and depression. Fang marks were observed in only two of the elk. One elk was bitten on two separate occasions.

A privately owned ocelot *(Felis pardalis)* was bitten on a paw by a northern Pacific rattlesnake that wandered into its cage. Signs were minimal, including swelling, erythema, and pain that was restricted to the bite site (Figure 24).

B. ELAPID (CORAL) SNAKES

1. Signs of Elapid Snakebite (Coral Snake) in Humans

Coral snakes must grasp a segment of anatomy, usually a digit, that is small enough to be surrounded by the open mouth and grasped while chewing. Pain and swelling at the bite site are usually minimal. Signs of neural dysfunction may begin within 90 min or be delayed for several hours. Signs include numbness and weakness of the affected limb, apprehension, drowsiness to unconsciousness, muscle fasciculation, tremors of the tongue, dysphagia, increased salivation, nausea, and vomiting.[6,8] Other signs reported include headache, photophobia, colic, miosis, dyspnea, convulsions, and paralysis. In fatal

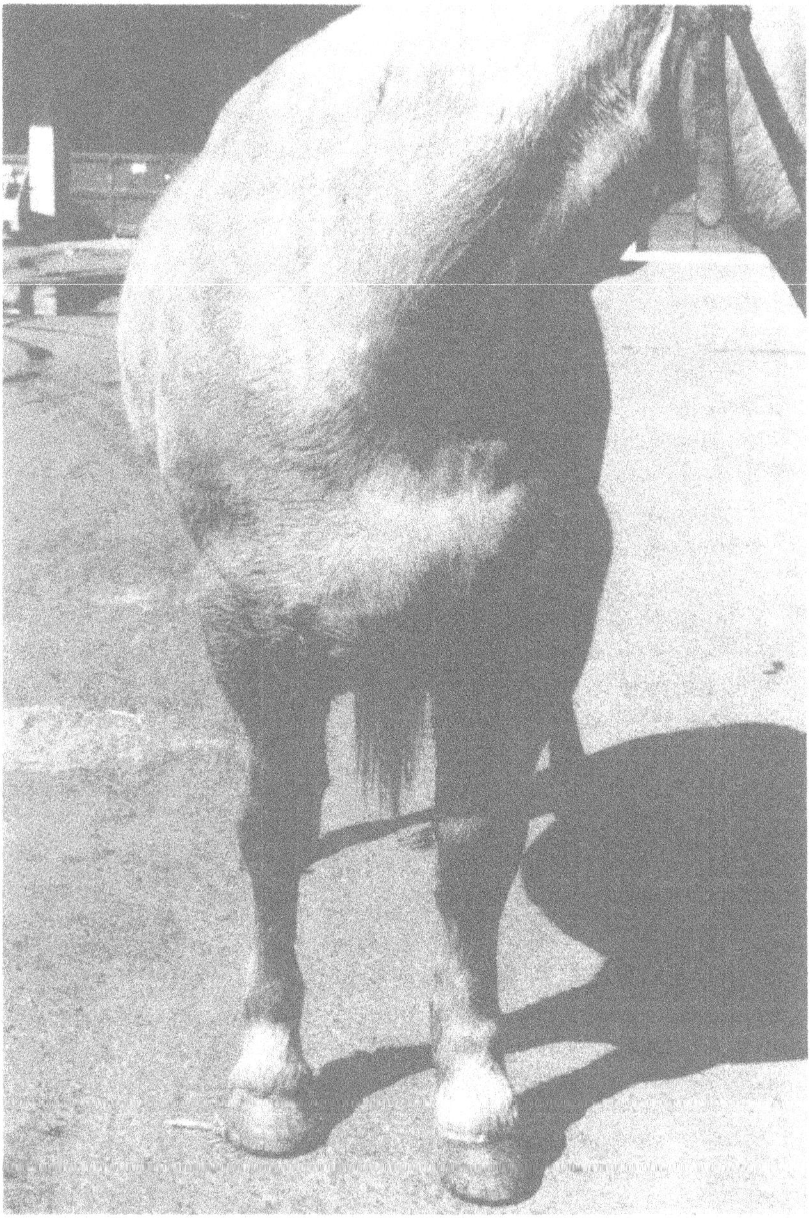

FIGURE 19. Progressive edema caused by a rattlesnake bite on the left carpus.

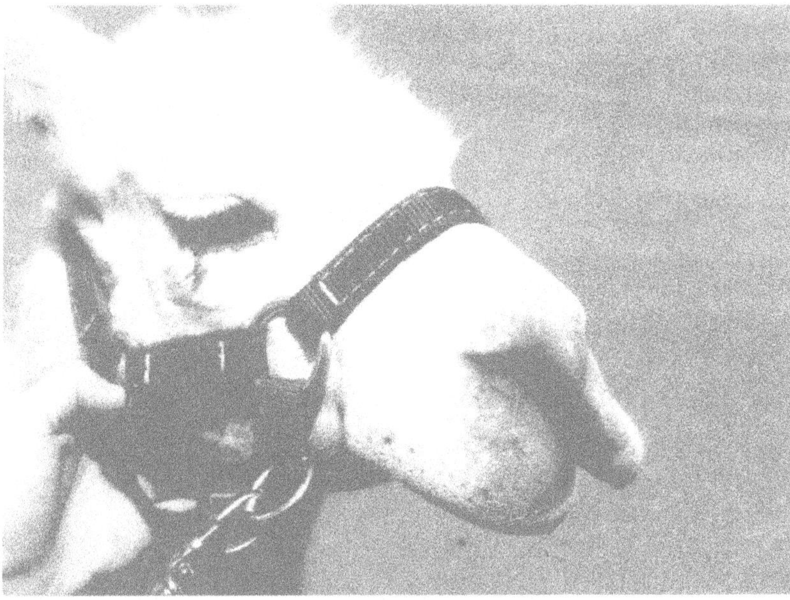

FIGURE 20. Swollen muzzle of a llama 24 h following a rattlesnake bite.

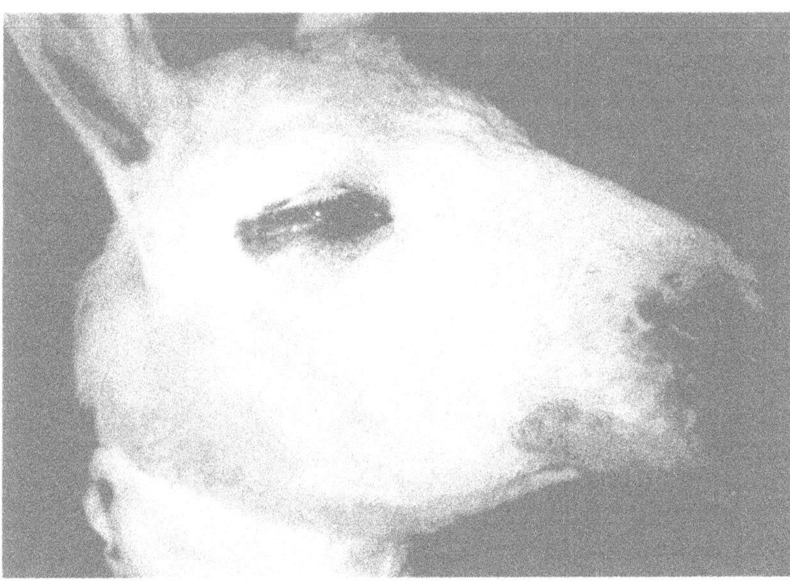

FIGURE 21. Swollen head of a llama 48 h after a rattlesnake bite.

FIGURE 22. Rattlesnake bite on the muzzle of a cow.

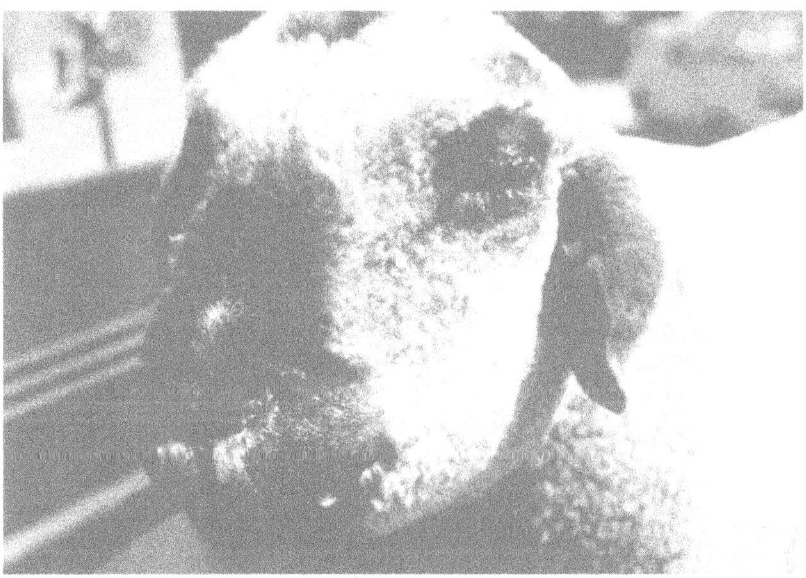

FIGURE 23. Rattlesnake bite on the head of a sheep.

FIGURE 24. Rattlesnake bite on the right paw of an ocelot, *Felis pardalis*.

bites, death usually results from paralysis of the respiratory musculature or cardiac failure.

2. Coral Snake Envenomation in Pets

Coral snakebite may occur in pets, but no descriptions of the clinical syndrome have been published. The dog has been used as an experimental subject for the study of coral snake venom. Signs reported from these studies indicate a syndrome similar to that of the human victim, with signs of muscular weakness, paresis, paralysis, dyspnea, dysphagia, muscle fasciculation, and cardiac failure.[24]

V. DIAGNOSIS OF ENVENOMATION

A history of bite exposure is confirmatory, but seldom available. Season of the year, prevalence of venomous snakes, and experience with previous snakebite cases are heavily relied upon. Evaluation of clinical signs is paramount.

A. LABORATORY SUPPORT

The toxic effects of crotalid snake venoms produce changes that are reflected in hematologic and biochemical values. Laboratory determinations may be used for initial evaluation of the case as well as for monitoring progress of therapy. The following laboratory analyses should be conducted in moderate to severe envenomation cases.

1. Basic hematology, including platelet count
2. Serum biochemistry, especially protein, serum urea nitrogen, creatinine, electrolytes, blood gases, and serum enzymes — particularly creatine phosphokinase (CK)
3. Coagulation studies, including fibrinogen, prothrombin, thrombin, clotting time, and platelet count

Common abnormalities are a reduction in erythrocyte numbers and hemoglobin concentration along with hypofibrinogenemia and thrombocytopenia. Clotting mechanisms are inhibited, resulting in prolonged prothrombin, thrombin, and clotting times. Serum enzyme levels may be elevated, with tissue necrosis. Excessive protein, glucose, and blood or hemoglobin are commonly observed in urine.

1. Serology

Recently, serologic tests (ELISA and immunodiffusion) have been used to identify snake venom antigens or antibodies.[23] Blood or serum oozing from a fang mark have been tested. In countries where numerous species of snakes are found, it is important to identify the snake so that specific antivenin may be administered. This is not of great consequence in crotalid bites in the U.S.

2. Differential Diagnosis

Small Pets: head trauma (contusion or cranial fractures), migrating grass awns, insect bites, and foreign body penetration must be considered.

Horse: snakebite is only one of several factors to consider in head swelling of a horse. Trauma and pharyngeal abscesses may obstruct venous drainage of the head. *Purpura hemorrhagica* (a toxic reaction to bacterial infection) produces similar swelling, but with considerable ecchymoses. Multiple stings from bees or wasps may also cause head swelling.

VI. TREATMENT OF CROTALID SNAKEBITE IN THE U.S.

For general management of crotalid snakebite in animals, refer to Chapter 14. Antivenin for crotalid bites in the U.S. is Wyeth's or Fort Dodge's *Antivenin (crotalidae) polyvalent* (Table 3 in Chapter 14).

A. HUMAN

Treatment for human crotalid snakebite in the U.S. is described in Chapter 14.[24-28]

B. DOG[9,11-14,29,30]

Severely envenomated dogs should be observed frequently. Periodic evaluation of blood pressure and an electrocardiogram is important during the course of therapy. If hemorrhage is already present when the patient is first examined, blood should be collected for cross-matching, in anticipation of the necessity for blood transfusion. Cross-matching cannot be performed once antivenin therapy begins.[1]

Administration of antivenin may be crucial to saving the life of a pet envenomated by a large eastern or western diamondback rattlesnake. The cost of therapy should be discussed with the client before administering antivenin because each vial costs from $130.00 to $150.00. The quantity of antivenin required for a dog is the same as for a human child or adult. The critical factor is the amount of venom injected that must be neutralized. Veterinarians practicing in areas inhabited by the diamondback rattlesnakes may administer five or more vials of antivenin.

The effectiveness of corticosteroid therapy is controversial, but corticosteroids are not contraindicated. Broad-spectrum antibiotics should be administered. It is not necessary to administer tetanus antitoxin or toxoid, as dogs are quite resistant to tetanus.[11]

Numerous other medications and forms of therapy for dogs have been reported in the literature (see Chapter 14). There are no reports of true anaphylactoid reactions to antivenin nor development of serum sickness in dogs.[9] Dogs do develop allergic reactions such as erythema of the pinnae that is responsive to benadryl administration.*

C. HORSE

Patency of the respiratory tract is the primary concern. If the bite is witnessed and the snake identified as venomous, a short segment of garden hose may be placed in a nostril to maintain patency before swelling obstructs the nares.[3,4] Such a tube may be sutured in place by a veterinarian. A more satisfactory method is placement of a tracheostomy tube, see page 132. Supplemental oxygen may be insufflated through a tracheostomy tube via a small tube from a tank of oxygen.

Broad-spectrum antibiotics should be administered along with tetanus antitoxin, or toxoid if vaccination is current. Horses that become hyperexcited may be quieted with xylazine hydrochloride (0.2 to 0.4 mg/kg) or acepromazine maleate (0.03 to 0.08 mg/kg). As both of these drugs cause lowering of the blood pressure, it is important to monitor the heart for a few minutes after administration.

Antivenin usage should be considered in envenomation of a foal or severe envenomation of an adult. However, economic factors may preclude this form of therapy.

D. LLAMA/ALPACA

Placement of a nasal tube may also be considered as first aid in observed bites of the llama or alpaca, but the space is much more narrow than in the horse and a correspondingly smaller tube is required. The smaller tube may not provide sufficient air flow. Tracheostomy, intravenous fluids to maintain hy-

* Personal communication, Dr. Michael Mount, Davis, CA.

dration, and time (swelling usually recedes sufficiently for the llama to breathe through the nostrils in 48 to 72 h) have been the author's primary method of therapy. When the swelling begins to recede, periodically occlude the tracheostomy tube to determine if the animal can breathe through the nostrils. The value of these animals warrants a discussion with the owner on the merits of antivenin therapy. If the bite or signs are observed and the animal is taken to a veterinary clinic within 2 h of the bite, the administration of antivenin may obviate the need to perform tracheostomy. Later administration of antivenin may shorten the course of envenomation.

The author found no mention in the literature of the effects of venom on a fetus of any animal. The epitheliochorial type of placentation of camelids may inhibit transfer of large-molecule toxins to the fetus, but this has not been determined experimentally. Surely, if severe hemolysis occurs from the effects of the venom, hypoxic death of the fetus may ensue. If a pregnant llama is bitten, antivenin therapy may minimize the possibility of fetal death caused by envenomation.

VII. TREATMENT OF CORAL SNAKEBITE IN DOGS

Coral snakes tend to bite and hold on; thus, pets may be observed with the snake still attached. In such cases, the pet should be transported immediately to a veterinary hospital and antivenin therapy begun at once. If therapy is delayed several hours until systemic signs develop, antivenin may be unable to reverse the effects of the neurotoxic venom. A specific antivenin, Wyeth's Antivenin *(Micrurus fulvius)*, is available for the eastern and Texas coral snakes, but, unfortunately, there is no protection against bites of the Sonoran coral snake, *Micruroides euryxanthus*.

If dyspnea indicates respiratory paralysis, positive pressure respiratory assistance may be necessary for hours to days. Other supportive and symptomatic therapy should also be given.

REFERENCES

1. **Wingert, W. A.,** Management of crotalid envenomations, in *Handbook of Natural Toxins, Vol. 5. Reptile Venoms and Toxins*, Tu, A. T., Ed., Marcel Dekker, New York, 1991, 611.
2. **Glenn, J. L. and Straight, R. C.,** The rattlesnakes and their venoms, yield and lethal toxicity, in *Rattlesnake Venoms*, Tu, A. T., Ed., Marcel Dekker, New York, 1982, 3.
3. **Burger, C. H.,** Snakebite, in *Equine Medicine and Surgery*, 2nd ed., Catcott, E. J., Ed., American Veterinary Publications, Santa Barbara, CA, 1972, 213.
4. **Burger, C. H. and Van Gelder, G. A.,** Snakebite, in *Equine Medicine and Surgery*, 3rd ed., Vol. 1, Mansmann, R. A., McAllister, E. S., and Pratt, P. W., Eds., American Veterinary Publications, Santa Barbara, CA, 1982, 215.
5. **Tu, A. T.,** Chemistry of rattlesnake venoms, in *Rattlesnake Venoms*, Tu, A. T., Ed., Marcel Dekker, New York, 1982, 247.
6. **Russell, F. E.,** *Snake Venom Poisoning*, Scholium International, Great Neck, NY, 1983.

7. **Minton, S. A., Jr. and Minton, M. R.,** *Venomous Reptiles,* Revised ed., Charles Scribner's Sons, New York, 1980.
8. **Parrish, H. M. and Khan, M. S.,** Bites by coral snakes: report of 11 representative cases, *Am. J. Med. Sci.,* 253, 561, 1967.
9. **Schaer, M.,** Eastern diamondback rattlesnake envenomation of 20 dogs, *Compendium of Continuing Education,* 6, 997, 1984.
10. **Mount, M. E.,** Toxins of animal origin, in *Textbook of Internal Veterinary Medicine,* 3rd ed., W. B. Saunders, Philadelphia, 1989, 472.
11. **Mansfield, P. D.,** The management of snake venom poisoning of dogs, *Compendium of Continuing Education,* 6, 988, 1984.
12. **Meerdink, G. L.,** Bites and stings of venomous animals, in *Current Veterinary Therapy, VII. Small Animal Practice,* Kirk, R. W., Ed., W. B. Saunders, Philadelphia, 1980, 174.
13. **Meerdink, G. L.,** Bites and stings of venomous animals, in *Current Veterinary Therapy, VIII. Small Animal Practice,* Kirk, R. W., Ed., W. B. Saunders, Philadelphia, 1983, 155.
14. **Oehme, F. W., Brown, J. F., and Fowler, M. E.,** Toxins of animal origin, in *Casarett and Doull's Toxicology — The Basic Science of Poisons,* 2nd ed., Doull, J., Klassen, C. D., and Amdur, M. O., Eds., Macmillan, New York, 1980, 557.
15. **Fowler, M. E.,** Snake envenomation, in *Medicine and Surgery of South American Camelids,* Fowler, M. E., Iowa State University Press, Ames, 1989, 379.
16. **Fowler, M. E.,** Zootoxins, in *Current Veterinary Therapy — Food Animal Practice 2,* Howard, J. L., Ed., W. B. Saunders, Philadelphia, 1986, 463.
17. **Fowler, M. E.,** Zootoxins, in *Plant Poisoning and Biotoxins,* Fowler, M. E., Reprographics, University of California, Davis, 1987, 158.
18. **Halmagyi, D. F. J., Starzecki, B., and Horner, G. J.,** Mechanism and pharmacology of shock due to rattlesnake venom in sheep, *J. Appl. Physiol.,* 20, 709, 1965.
19. **Mount, M.,** Venomous animal-related intoxications, in *Large Animal Internal Medicine,* Smith, B. P., Ed., C. V. Mosby, St. Louis, MO, 1990, 1662.
20. **Saliba, A. M., Belluomini, H. E., and Leinz, F. F.,** Experimentelle Crotalus-Vergiftung bei Rindern — Anatomisch-pathologische studie (Experimental rattlesnake venom poisoning in cattle, an anatomic and pathologic study), *Deutsche Tierarztliche Wochenschrift,* 90, 503, 1983.
21. **Blood, D. C. and Radostits, O. M.,** *Veterinary Medicine,* 7th ed., Bailliere Tindall, London, 1989, 1356.
22. **Fowler, M. E.,** Toxicities in exotic and zoo animals, *Vet. Clin. N. Am.,* 5, 685, 1975.
23. **Miller, M. W., Wild, M. A., Baker, B. J., and Tu, A. T.,** Snakebite in captive Rocky Mountain elk *(Cervus elaphus nelsoni), J. Wildlife Dis.,* 25, 392, 1989.
24. **Russell, F. E.,** Snake venom poisoning, *Vet. Human Toxicol.,* 33, 584, 1991.
25. **Arnold, R. E.,** Treatment of rattlesnake bites, in *Rattlesnake Venoms,* Tu, A. T., Ed., Marcel Dekker, New York, 1982, 315.
26. **Glass, T. G., Jr.,** Management of the western diamondback rattlesnake bite, in *Rattlesnake Venoms,* Tu, A. T., Ed., Marcel Dekker, New York, 1982, 339.
27. **Moore, G. M., Ed.,** *Poisonous Snakes of the World,* NAVMED P - 5099, Revised 1965, Dept. of the Navy, Bureau Medicine and Surgery, U.S. Government Printing Office, Washington, D.C., 1965.
28. **Russell, F. E. and Saunders, P. R.,** *Animal Toxins,* Pergamon Press, New York, 1966.
29. **Parrish, H. M. and Scatterday, J. E.,** A survey of poisonous snakebites among domestic animals in Florida, *Vet. Med.,* 52, 135, 1957.
30. **Parrish, D. M., Scatterday, J. E., and Pollard, C. B.,** The clinical management of snake venom poisoning in domestic animals, *J. Am. Vet. Med. Assoc.,* 130, 548, 1957.
31. **Campbell, J. A. and Lamar, W. W.,** *The Venomous Reptiles of Latin America,* Comstock Publishing, Ithaca, NY, 1989.

Chapter 16

SNAKEBITE IN MEXICO AND CENTRAL AND SOUTH AMERICA

I. VENOMOUS SNAKES

New World areas south of the U.S. border support a rich snake fauna. Numerous venomous species are known, from both the *Elapidae* and *Crotalidae* families.[1-3] Fortunately, many species do not present a serious hazard to humans or animals because of small size, only mildly toxic venom, or a limited distribution. Table 1 lists some of the dangerous species. Table 2 lists relative toxicities.

Coral snakes are the only elapids. The genus *Micrurus* is broadly distributed. Of the 53 species that have been identified in Mexico and Central and South America, many are small, shy, and of little consequence in human or animal envenomation. Others, such as the Brazilian giant coral (southern coral) snake *(Micrurus frontalis)* may pose a threat to animals. *M. frontalis* averages 70 to 90 cm (28 to 36 in.) in length. It is distributed from southern Brazil west to eastern Bolivia, and south through Paraguay, Uruguay, and north central Argentina.[1] The range of the Sonoran coral snake *(Micruroides euryxanthus)* extends from Arizona into northern Mexico. No instance of animal poisoning from this species has been reported in the literature.

Dangerous pit vipers produce significant morbidity and mortality in both animals and humans (Table 1). Though 24 species of rattlesnake *Crotalus* spp. inhabit Mexico, only 3 species account for the majority of lethal envenomations.

Northern Mexico is included in the range of the western diamondback rattlesnake, *Crotalus atrox* (Figure 1). See Chapter 15 for further description. *C. scutulatus*, commonly called the Mojave or green rattlesnake in the U.S., inhabits only a limited area of the Mojave desert of southern California. It has a broader distribution throughout the Chihuahuan Desert of northern Mexico (See Figures 5 and 6 in Chapter 15). See Chapter 15 for a description of this snake and clinical signs of envenomation from its bite.[1,2]

The neotropical rattlesnake, *Crotalus durissus*, is unique to the region described in this chapter. The neotropical rattlesnake is heavy bodied, commonly reaching 100 cm (40 in.) in length. Large males may grow to twice that length (180 cm, 71 in.). Neotropical rattlesnakes are quick to assume a dramatic defensive coil (Figure 2) and strike viciously.[1] The aggressive behavior and large size combine with production of a toxic venom to make this snake extremely dangerous.

The widespread southern subspecies *Crotalus durissus terrificus* (Figure 2) has a patchy distribution in South America. Though absent from most of the Amazon Basin, it occupies habitats in central and southern Brazil, westward into Bolivia and Peru, and south to Paraguay, Uruguay, and northern Argentina (Figure 3).[1] This subspecies produces a highly toxic neurotoxin.

TABLE 1
Selected Venomous Snakes — Mexico, Central and South America

Phylum — *Chordata*
Class — *Reptilia*
Order — *Squamata*
Suborder — *Ophidia (Serpentes)* — snakes
  Family — *Elapidae*
    Micruroides euryxanthus — Sonoran coral snake
    Micrurus fulvius fulvius — eastern coral snake
    Micrurus frontalis — Brazilian giant coral snake
    Micrurus spp. — coral snakes, 53 species in North, Central, and South America
  Family — *Crotalidae* — pit vipers
    Bothrops asper — fer-de-lance (terciopelo)
    B. atrox — fer-de-lance (barba amarilla, caissaca)
    B. jararaca
    B. jararacussu — yarara cussu
    Crotalus durissus durissus — Central American rattlesnake (RS)
    C. durissus terrificus — South American RS, cascabel
    C. atrox — western diamondback rattlesnake
    C. scutulatus — Mojave (green) rattlesnake
    Agkistrodon bilineatus — cantil
    Lachesis mutus mutus — bushmaster

TABLE 2
Relative Toxicity of Selected Latin American Snake Venoms

Common and scientific names	LD_{50} dose in mice, i.v. (mg/kg)	Venom yield, dry (mg)	Approximate length (cm)
Bushmaster			
Lachesis muta	4.5		152–213
Common lancehead			
Bothrops atrox	1.4		122–183
Fer-de-lance			
Bothrops asper	1.2		
South American rattlesnake	0.13–0.35	24–44	153
Crotalus durissus terrificus			
Mojave rattlesnake	0.1–0.2	50–90	56–102
Crotalus scutulatus			
Western diamondback rattlesnake	4.2	77–423	60–168
Crotalus atrox			
Brazilian giant coral snake	0.63		70–90
Micrurus frontalis			

Data from Kohle, V., *Handbook of Natural Toxins, Vol. 5. Reptile Venoms and Toxins,* Marcel Dekker, New York, 1991.

FIGURE 1. Distribution of western diamondback rattlesnake, *Crotalus atrox*, in North America. (Modified from Campbell, J. A. and Lamar, W. W., *The Venomous Reptiles of Latin America*, Comstock Publishing, Ithaca, NY, 1989. With permission.)

FIGURE 2. Neotropical rattlesnake, *Crotalus durissus terrificus*.

FIGURE 3. Distribution of neotropical rattlesnake, *Crotalus durissus terrificus*, in South America. (Modified from Campbell, J. A. and Lamar, W. W., *The Venomous Reptiles of Latin America*, Comstock Publishing, Ithaca, NY, 1989. With permission.)

The Central American rattlesnake, *Crotalus durissus durissus* (cascabel), is the northern subspecies of the neotropical rattlesnake. Its distribution includes the east and west coasts of southern Mexico into Central America (Figure 4). The cascabel is depicted on many Mayan ruins.

The cantil, *Agkistrodon bilineatus*, has a limited distribution on the west coast of Mexico, south to Costa Rica (Figure 5).[1] Although fatal bites of humans have been reported, no information is available regarding animal bites. The cantil is a heavy-bodied snake with a long slender tail and may reach 1.38 m (54 in.) in length. The fangs are relatively longer than those of any other North American snake of its size.[1] It is aggressive and may bite repeatedly.

The genus *Bothrops*, South American lanceheads, is comprised of 31 species distributed primarily in South America, but *Bothrops asper* extends through Central America and along the east coast of Mexico (Figure 6). The four most dangerous species are listed in Table 1. The name "fer-de-lance" is commonly applied to *B. asper* and *B. atrox* in the English literature, but many Spanish or Indian names are used locally.[1]

Bothrops asper is a large snake, some reaching lengths in excess of 2.5 m (8.2 ft). Body colors and patterns vary with the locality. *B. atrox* is generally

FIGURE 4. Distribution of neotropical rattlesnake, *Crotalus durissus durissus*, in Mexico and Central America (Modified from Campbell, J. A. and Lamar, W. W., *The Venomous Reptiles of Latin America*, Comstock Publishing, Ithaca, NY, 1989. With permission.)

smaller than *B. asper*, rarely reaching 1.5 m (5 ft) in length. Most individuals are no longer than 1.0 m (3.2 ft). Two other large species, *B. jararaca* and *B. jararacussu* (Figure 7), along with the previous two species are responsible for considerable human morbidity and mortality in Central and South America. Distribution of selected *Bothrops* spp. is illustrated in Figure 6. Snakebite in animals is not well documented, but the pattern is probably similar to that of human envenomation.

The bushmaster, *Lachesis muta*, ranges from Nicaragua to northern South America (Figure 8). This is a huge snake, with many individuals reaching 2.0 m (7 ft) in length, and exceptional specimens growing as long as 3.6 m (11.8 ft). Bushmasters are nocturnal and aggressive if disturbed.[1]

II. CONDITIONS OF ENVENOMATION

Each snake species has a preferred habitat and animals cohabiting the same environment may be at risk. The basic conditions for envenomation are the same as for the snakes of the U.S. (see page 140, Chapter 15).

FIGURE 5. Distribution of cantil, *Agkistrodon bilineatus*, in Mexico and Central America. (Modified from Campbell, J. A. and Lamar, W. W., *The Venomous Reptiles of Latin America*, Comstock Publishing, Ithaca, NY, 1989. With permission.)

III. VENOM[4-8]

All the venoms of the pit vipers of the New World are similar.[2] Differences in toxicity are determined by the injected dose and other factors as described on page 113 of Chapter 13. A unique crotalid venin, crotoxin, is found in the venom produced by the South American rattlesnake, *Crotalus durissus terrificus*. Crotoxin is a potent, presynaptic neurotoxin, or group of neurotoxins.[4] For many years it was thought that this and the Mojave rattlesnake (Mojave toxin) were the only crotalids that produce neurotoxic venoms. Recent investigations indicate that venoms of a number of species contain small amounts of neurotoxins, but actions of other venins overshadow those of the neurotoxin.

There are geographical differences in the potency of neurotoxins.[4] Both of the crotalid neurotoxins also produce a myonecrotic effect.[4]

IV. CLINICAL SIGNS

A. HUMAN SNAKEBITE[9]

Coral snake venom is primarily neurotoxic, producing a progressive skeletal muscle paralysis. As in other elapid envenomations, the onset of signs may be delayed for 2 to 6 h after the bite. Maximum effect may not be reached for as long as 48 h. Human victims experience double vision, ptosis, oculoplegia,

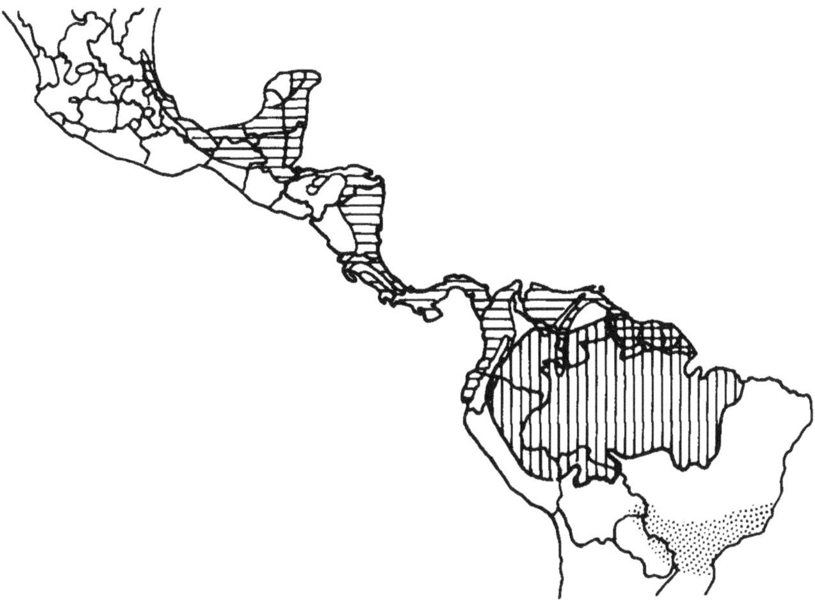

FIGURE 6. Distribution of: ▤ Fer-de-lance, *Bothrops asper*. ▥ Fer-de-lance, *Bothrops atrox*. ▩ Yarara cussu, *Bothrops jararacussu*. (Modified from Campbell, J. A. and Lamar, W. W., *The Venomous Reptiles of Latin America*, Comstock Publishing, Ithaca, NY, 1989. With permission.)

painful joints, dysphagia, and generalized muscle weakness. Without respiratory assistance, paralysis of the respiratory muscles may cause death.[1]

Human bites from the nonneurotoxic crotalids *(Crotalus, Agkistrodon, Bothrops,* and *Bothriopsis)* are similar to crotalid snakebite in the U.S. (see Chapter 15, page 149).

Bites of the neotropical rattlesnake, *Crotalus durissus*, are highly dangerous, especially those of the South American subspecies *C. d. terrificus*. Signs of neurotoxicity predominate, with progressive paralysis of cranial and peripheral nerves. Within 30 to 60 min of the bite, ptosis is observed along with inability to move the eyeball. Vestibular paralysis results in incoordination. Mydriasis is a sign of severe and possibly fatal envenomation.[9] Dysphagia results in excessive salivation. Progressive, peripheral nerve paralysis results in paresis, incoordination, dyspnea, and ultimately, respiratory paralysis.

The venoms of South American rattlesnakes also cause variable degrees of hemolysis. The released hemoglobin may impact the renal tubules, resulting in anuria. A direct nephrotoxin may also be involved.[9] Hypovolemic shock may ensue from hemolysis and kidney malfunction.

FIGURE 7. Yarara cussu, *Bothrops jararacussu*.

FIGURE 8. Distribution of the bushmaster, *Lachesis muta*. (Modified from Campbell, J. A. and Lamar, W. W., *The Venomous Reptiles of Latin America*, Comstock Publishing, Ithaca, NY, 1989. With permission.)

TABLE 3
Clinical Syndromes of Selected Latin American Venomous Snakebites

Toxic action	Proteolytic and coagulant	Neurotoxic and hemolytic	Neurotoxic
Species involved	*Crotalus* spp., *Agkistrodon* spp., *Bothrops* spp., *Bothriopsis* spp., *Lachesis muta*	*Crotalus scutulatus*, *Crotalus durissus terrificus*, South American rattlesnake	*Micrurus* spp., coral snakes
Local edema/necrosis	+3	0 – +1	0
Cranial nerve paralysis, ptosis, mydriasis, occuloplegia	0	+2	+3
Peripheral nerve paralysis	0	+3	+4
Hemoglobinuria	0	+2	0
Hematuria	+1	0	0
Anticoagulant	+3	0 – +2	0
Cause of death	Necrotizing effects on kidney, liver, heart, lung	Respiratory paralysis, renal malfunction	Respiratory paralysis
Treatment	Antivenin plus local and supportive	Early antivenin	Early antivenin

B. ANIMAL ENVENOMATION[10,11]

Few clinical reports describe syndromes of animal envenomation in Mexico or Central or South America. However, some reports describe the experimental production of envenomation in laboratory animals, dogs, and livestock species.[11,12] All mammals are susceptible to the venoms. The effect is somewhat variable, but variation may depend on victim size rather than innate resistance. The syndromes produced are likely to be similar to those of human envenomation.[9]

Clinical signs reported during experimental poisoning of cattle included ataxia, falling, recumbency (sternal or lateral), reduced palpebral reflex but no ptosis, and paralysis of ocular muscles. Other less-consistent signs included depression, salivation, anorexia, thirst, rumen stasis, and bloat. It was reported that cattle are extremely sensitive to the neurotoxin and less sensitive to the hemolytic fraction. Death usually occurred within 24 h.[11]

V. DIAGNOSIS

Three basic syndromes are outlined in Table 3. Necropsy lesions from animal envenomation by snakes producing a venom with proteolytic and/or

TABLE 4
Recommended Antivenins for Treating Venomous Snakebite in Latin America

Common name	Scientific name	Antivenin[a]
Coral snakes	*Micrurus fulvius*, *Micrurus* spp.	Wyeth monovalent antivenin, *Micrurus fulvius*, or Antielapidico[b]
Neotropical rattlesnake	*Crotalus durissus durissus*	Anticrotalico monovalent
South American rattlesnake	*C. durissus terrificus*	
Mojave rattlesnake	*Crotalus scutulatus*	Wyeth polyvalent crotalidae
Western diamondback rattlesnake,	*Crotalus atrox*	Wyeth polyvalent crotalidae
Rattlesnakes general	*Crotalus* spp.	
Fer-de-lance	*Bothrops asper*, *Bothrops atrox*	Antibotropico monovalent or polyvalent
Yarara cussu	*Bothrops jararacussu*, *Bothrops jararaca*	
Bushmaster	*Lachesis muta*	Antilaquético, monovalent

[a] See Table 3, Chapter 14 for more details.
[b] Produced at Instituto Butantan, Brazil. Similar antivenins produced in other Latin American countries.

coagulant action may include hepatic necrosis, renal necrosis, cardiomyopathy, disseminated intravascular coagulation, and edema and necrosis at the bite site. Neurotoxins do not produce organic lesions, but hemolysis causes anemia and nephrosis.

Lesions observed at necropsy in experimental envenomation of cattle included general hemorrhage of parenchymal organs, but most pronounced in the central nervous system, arteriolar hyaline necrosis of the brain and lungs, focal glomerulonephritis, tubular necrosis, hemorrhagic edema at the bite site, purulent myositis, degenerative hepatopathy, and myocardiopathy.[13]

VI. TREATMENT

A. HUMAN

Antivenin, administered as quickly as possible after envenomation is assured, is the treatment of choice, Table 4. Antivenins are produced in Latin American countries where human snakebite is a significant problem. Victims of bites from neurotoxic snakes require close observation and use of assisted respiration in the event of respiratory paralysis. Antivenin administered after paralysis is advanced will neutralize any circulating venom, but may not reverse neural dysfunction. Neurotoxins produce an irreversible biochemical reaction. Generation of nonaffected molecules is required to restore neural function. Support therapy for the victim must be maintained until function is restored.[9]

B. ANIMAL

Antivenins are the treatment of choice in animal bites also, but the cost of antivenin therapy may be prohibitive.[12,13] In Brazil, there have been times when it was illegal to administer antivenin to animals because all available antivenin was needed for treatment of human snakebite victims.

Animal snakebite in Latin America is treated similarly to bites of comparable species in the U.S. (see Chapters 14 and 15).

REFERENCES

1. **Campbell, J. A. and Lamar, W. W.,** *The Venomous Reptiles of Latin America*, Comstock Publishing, Ithaca, NY, 1989.
2. **Glenn, J. L. and Straight, R. C.,** The rattlesnakes and their venom yield and lethal toxicity, in *Rattlesnake Venoms*, Tu, A. T., Ed., Marcel Dekker, New York, 1982, 3.
3. **Hoge, A. R., Alma, S., and Romano, R. W. D. L.,** Neotropical pit vipers, sea snakes and coral snakes, in *Venomous Animals and Their Venoms*, Bucherl, W. and Buckley, E. E., Eds., Academic Press, Orlando, FL, 1971, 211.
4. **Hendon, R. A. and Bieber, A. L.,** Presynaptic toxins from rattlesnake venoms, in *Rattlesnake Venoms*, Tu, A. T., Ed., Marcel Dekker, New York, 1982, 211.
5. **Kaiser, E. and Michl, H.,** Chemistry and pharmacology of the venoms of *Bothrops* and *Lachesis*, in *Venomous Animals and Their Venoms*, Bucherl, W. and Buckley, E. E., Eds., Academic Press, Orlando, FL, 1971, 308.
6. **Kohle, V.,** Toxicity of snake venoms and their components, in *Handbook of Natural Toxins, Vol. 5. Reptile Venoms and Toxins*, Marcel Dekker, New York, 1991, 405.
7. **Tu, A. T.,** Chemistry of rattlesnake venoms, in *Rattlesnake Venoms*, Tu, A. T., Ed., Marcel Dekker, New York, 1982, 247.
8. **Tu, A. T.,** Local tissue damaging (hemorrhage and myonecrosis) toxins from rattlesnake and pit viper venoms, *J. Toxicol. Toxin Rev.*, 205, 91, 1983.
9. **Rosenfeld, G.,** Symptomatology, pathology and treatment of snake bites in South America, in *Venomous Animals and Their Venoms*, Bucherl, W. and Buckley, E. E., Eds., Academic Press, Orlando, FL, 1971, 346.
10. **Grunet, E.,** Snakebite wounds in large animals in Brazil, *Deutsch Tierärztl. Wschr.*, 74, 31, 1968.
11. **Belluomini, H. E., Aranjo, P., Rosenfeld, G., Leinz, F. F., and Birgel, E. H.,** Symptomstologie der experimentellen Crotalustoxin-Vergiftung bei Rindern die einer spezifischen Serumtherapie unterworfen wurden (Symptomatology of experimental crotalid envenomation in cattle being submitted to a specific sero-therapy), *Dtsch. Tierärztl. Wschr.*, 89, 433, 1982.
12. **Belluomini, V. E., Aranjo, P., Rosenfeld, G., and Penha, A. M.,** Eintrag zur Serumtherapie bei experimentellen Vergiftung von Rindern mit dem gifte der Klopperschlange (Report on antivenin treatment in experimental rattlesnake poisoning in cattle), *Dtsch. Tierärztl. Wschr.*, 90, 81, 1983.
13. **Saliba, A. M., Belluomini, H. E., and Leinz, F. F.,** Experimentelle Crotalus-Vergiftung bei Rindern-Anatomisch-pathologische Studie (Experimental poisoning with crotalid venom in cattle — an anatomic pathologic study), *Dtsch. Tierärztl. Wschr.*, 90, 503, 1983.

Chapter 17

SNAKEBITE IN AUSTRALIA

I. VENOMOUS SNAKES

Five families of terrestrial snakes with 36 genera and 136 described species are found in Australia.[1] More than half of Australian snakes are venomous, members of the *Elapidae* family which includes 22 genera and 80 species. Fortunately, many venomous species are small, have a mildly toxic venom, or have a narrow distribution. Dangerous snakes are limited to a few major species (See Table 1).

Sea snakes of the family *Hydrophoiidae* are also found in the western, northern, and eastern shallow coastal waters. Sea snakes are not unique to Australia, and are of little consequence in animal envenomation. However, a short discussion of human sea snake envenomation is included in this chapter in order to include all venomous snake groups.

A. DESCRIPTION OF SELECTED SPECIES[1-3]

The tiger snake, *Notechis scutatus* (Figure 1), is notorious as a snake dangerous to humans and animals throughout its range, (Figure 2). Its average length is 1.1 m (3.5 ft), but individuals may reach a length of 1.8 m (6 ft). The basic body color varies from light gray to dark or greenish brown, usually with dark transverse bands.

The death adder, *Acanthophis antarcticus,* is a smaller, stout-bodied snake with lengths up to 0.9 m (3 ft). It has a broad head and a banded tail. Its distribution is illustrated in Figure 3.

The common taipan, *Oxyuranus scutellatus,* is the largest [up to 3.4 m (11 ft)] and most dangerous snake in Australia, with a distribution from Grafton in northern New South Wales, along the Queensland coast, and in the Darwin area (Figure 4). In the interior, Cooper Creek, Diamentina, and Georina River basins, and further down in the Murray-Darling River systems, the western taipan, inland taipan, or fierce snake has been identified as a distinct species. Taipan venom is recognized as the most toxic land snake venom known. Both snakes have a basic body color of dark brown with a yellowish belly. The head is large and there is a distinct neck.

There is confusion over the term "brown snake" because many Australian snakes have a brownish color in one phase or another. However, three species are classified as brown snakes by herpetologists. The common or eastern brown snake, *Pseudonaja textilis*, is the most common dangerous snake encountered by humans and animals in the eastern half of the country (Figure 5). The dugite, *Pseudonaja affinis,* has a limited distribution in southwestern Australia (Figure 6) and the gwardar *(Pseudonaja nuchalis)* is distributed over much of the country except for southern and eastern coastal areas (Figure 6). Brown snakes are slim, swift snakes with lengths averaging 1.2 m. Some

TABLE 1
Venomous Australian Snakes

Phylum — *Chordata*
Class — *Reptilia*
Order — *Squamata*
Suborder — *Ophidia (Serpentes)* — snakes
  Family — *Elapidae*
    Oxyuranus scutellatus — taipan
    O. microlepidotus — inland taipan snake (small-scaled snake, western taipan, "fierce snake")
    Pseudechis australis — mulga (king brown) snake
    P. colletti — Collett's snake
    P. porphyriacus — red-bellied black snake
    P. guttatus — spotted black (blue-bellied black) snake
    Acanthophis antarcticus — common death adder
    Pseudonaja textilis — common brown (eastern brown) snake
    P. affinis — dugite
    P. nuchalis — gwardar (western brown) snake
    Notechis ater — black tiger snake (4 subspecies)
    N. scutatus — eastern tiger snake (2 subspecies)
    Austrelaps superba — copperhead
    Tropidechis carinatus — rough-scaled (Clarence river snake)
    Cryptophis nigrescens — eastern small-eyed snake

FIGURE 1. Tiger snake, *Notechis scutatus.*

FIGURE 2. Distribution of the tiger snake, *Notechis scutatus* in Australia. (Modified from Wilson, S. K. and Knowles, D. G., *Australia's Reptiles*, Collins, Sydney, Australia, 1988. With permission.)

individuals reach lengths of 1.8 m (6 ft). They are nervous, alert, and quick to strike, but fortunately have short fangs which may be unsuccessful in penetrating the thick hairy skin of adult livestock. Nevertheless, the common brown and the tiger snake account for the majority of animal bites in Australia.

The copperhead, *Austrelaps superba,* is not related to the North American copperhead. It is a sluggish, thick-set, yellow-brown snake reaching a length of 1.6 m (5.5 ft). Its distribution is limited to southeastern Australia and Tasmania (Figure 5).

The black snake group contains two dangerous Australian species. The king brown or mulga black snake, *Pseudechis australis,* is actually a coppery-brown to tan color and is found throughout much of Australia, except the south and southeast coastal areas (Figure 7). The red-bellied black snake, *Pseudechis polyphyriacus,* is found east of the dividing range, into Victoria, New South Wales, and South Australia (Figure 4). A New Guinea species, the Papuan

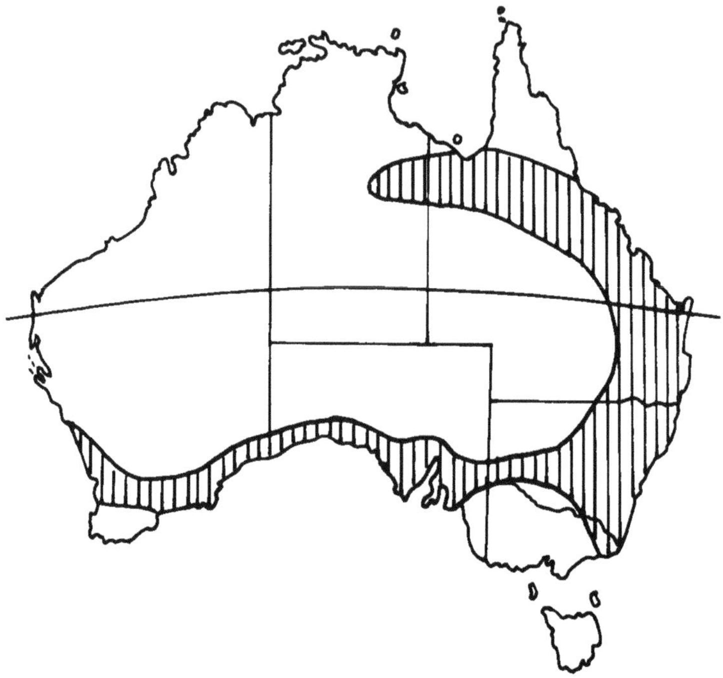

FIGURE 3. Distribution of the death adder, *Acanthophis antarcticus*. (Modified from Wilson, S. K. and Knowles, D. G., *Australia's Reptiles*, Collins, Sydney, Australia, 1988. With permission.)

black snake *(Pseudechis papuanus)* inhabits the entire coast of Papua, New Guinea.

At least 10 species of sea snakes may be found in the waters surrounding Australia (Figure 8), the most commonly encountered being *Enhydrina schistosa*.[4-6] A number of other genera are represented *(Hydrophis, Lapemis,* and *Laticauda)*, none of which are of consequence in domestic animal poisoning. Their effect on marine animals, other than prey species, is unknown.

II. CONDITIONS OF ENVENOMATION

Dangerous venomous snakes live throughout Australia. Most snakes tend to avoid areas with high human impact; in a few areas, humans may have eradicated snakes, but there are ample opportunities for human and animal interaction with venomous snakes. Although some species are pugnacious when cornered or harassed, most will avoid confrontation if possible. Snakebite rarely occurs during the winter months of May through August in the more temperate regions of the country, but snakes are active year-round in the northern tropical regions.

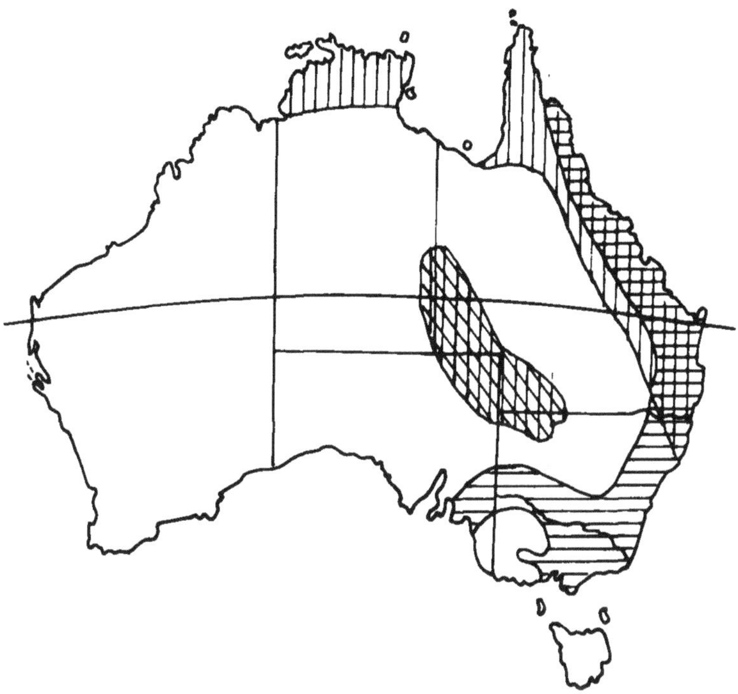

FIGURE 4. Distribution of: ▥ Taipan, *Oxyuranus scutellatus*. ▩ Inland taipan, *O. microlepidotus*. ▤ Red-bellied black snake, *Pseudechis porphyriacus*. (Modified from Wilson, S. K. and Knowles, D. G., *Australia's Reptiles*, Collins, Sydney, Australia, 1988. With permission.)

A. HORSES

Head, muzzle, face, and neck bites are most common in foals from a few days to 50 d of age.[7] Foals lack experience in dealing with potential threats and have an innate curiosity that leads them to investigate strange creatures in their paddocks or pastures. The hair and thickness of adult horse skin may minimize the risk of a snake's fangs penetrating to the subcutis, but a large, agitated snake may envenomate if encountered at close quarters.

Unique environmental conditions may contribute to the prevalence of snake bite in a given area. In Queensland, the heavy black soil develops large cracks that provide refuge for snakes during the heat of the day. When rains come, the snakes are driven out of the cracks.[7] On one racing stud farm, mares were fed supplemental alfalfa hay on the ground and snakes were known to shelter beneath the piles of hay during rain storms. Horses could easily disturb a snake while attempting to eat the hay.[7]

B. DOGS

Australian snakebite literature is replete with references to snake killer dogs that get bitten. Aggressive dogs may attempt to kill any snake encountered and

FIGURE 5. Distribution of: ▤ Common brown snake, *Pseudonaja textilis*. ▥ Copperhead, *Austrelaps superba*. (Modified from Wilson, S. K. and Knowles, D. G., *Australia's Reptiles*, Collins, Sydney, Australia, 1988. With permission.)

be bitten in the process. Pets are at the same risk as their owners for exposure to venomous snakes.

III. VENOM CHARACTERISTICS[8,9]

All the dangerous land snakes in Australia are in the family *Elapidae* and their venoms have a predominantly neurotoxic action. There are marked differences in the neurotoxic action of venoms from various species.[10-13] The venom of the tiger snake, death adder, copperhead, brown snake, and taipan are highly neurotoxic, while venoms of the black snakes are milder in effect. The neurotoxin of the common brown snake differs antigenically from neurotoxins of the tiger snake or taipan and corresponding antivenins* are not mutually protective. See Chapter 13 for a more detailed discussion of neurotoxins.

There is frequently a latent phase following a bite until signs of motor paralysis begin, but the progress of the paralysis may be rapid and fulminating shortly thereafter. In dogs, an interesting reported reaction is peracute collapse within minutes of the bite, from which the dog recovers for a time, up to hours, after which paralytic signs reappear, progressing to death if the dog is un-

* The preferred spelling for antivenin in Australia is antivenene.

FIGURE 6. Distribution of: [vertical hatching] Gwardar, *Pseudonaja nuchalis*. [horizontal hatching] Dugite, *Pseudonaja affinis affinis*. (Modified from Wilson, S. K. and Knowles, D. G., *Australia's Reptiles*, Collins, Sydney, Australia, 1988. With permission.)

treated.[1] This effect could be related to an intravenous envenomation or envenomation into a highly vascular region.

Australian snake venoms also may have other actions (Table 2), including interference with the clotting mechanism.[14-16] Common brown, taipan, and tiger snake venoms activate the production of thrombin which acts on fibrinogen to produce fibrin (the basis for clotting). However, instead of producing generalized intravascular clotting, the clinical effect is to defibrinate the blood, enhancing hemorrhage, which is particularly evident in the dog.[1] Within 30 min of a lethal tiger snake bite in a dog, hemorrhage from the bite wound may resume and needle pricks from administration of medication may fail to clot.[23] Additional signs associated with hemorrhage may include hematemesis, hematuria, and melena. The venoms of other Australian snakes have no or minimal coagulant components.

Venoms may contain venins that have direct hemolytic action or hemolysis may occur secondarily to the action of phospholipase A_2. Lysis of erythrocytes may cause icterus and hemoglobinuria, which accompany bites of the taipan and red-bellied black snakes.[1] A cytotoxic component of some venoms may contribute to total hemorrhage by damaging vascular endothelium, plus destroying muscle cells, renal epithelium, and platelets.[1]

FIGURE 7. Distribution of: ▥Mulga or king brown snake, *Pseudechis australis*. ▤Rough-scaled snake, *Tropidechis carinatus*. (Modified from Wilson, S. K. and Knowles, D. G., *Australia's Reptiles*, Collins, Sydney, Australia, 1988. With permission.)

Table 2 summarizes the quantity of venom produced in an average milking, and shows toxicity of the venom and the relative importance of the major toxic actions of the venoms of some of the dangerous species. Table 3 illustrates the variation in toxicity of the venom of the tiger snake in victim species. Note that tiger snake venom is 100 × more toxic to the horse than to the rat.

The venom of sea snakes is primarily neurotoxic. Human victims succumb rapidly to respiratory paralysis.

IV. CLINICAL SIGNS

Veterinarians dealing with snakebite in Australia see a variety of clinical signs and frequently are unable to be certain which species of snake bit the animal.[26] All venoms produce some degree of neurotoxicity, but the severity of signs and progress of the syndrome vary with the amount of venom injected, the species of snake, and the age, size, and species of the victim.[1] The syndromes reported by various authors are dependent on their location in Australia and the snakes found in that region; however, a common clinical sign of a lethal bite is an ascending paralysis beginning with the hind quarters and moving cranially, ultimately paralyzing respiration. Paralysis of cranial nerves devel-

FIGURE 8. World distribution of sea snakes. (Modified from Barme, M., *Venomous Animals and Their Venoms, Vol. 1. Venomous Vertebrates*, Bucherl, W., Buckley, E. E., and Deulofeu, V., Eds., Academic Press, Orlando, FL, 1968. With permission.)

ops early in the course of the syndrome and includes immobilization of the ciliary body (mydriasis, no response to light), inability to swallow, and inability to vocalize.

A. DOG[14-17,19-24]

Signs upon presentation vary with the time elapsed since the bite. Cytotoxic effects at the bite site are absent or minimal in most Australian snakebites. Snake venoms having hemolytic and coagulant actions may produce a delayed effect at the bite site causing slight edema, local hemorrhage, and oozing of blood from the bite wounds. A dog that has fought with a snake may be exhausted from the encounter or the initial effects of the venom may give the appearance of exhaustion. Even quiet dogs tend to become hyperexcited and may attempt to bite when handled for examination or treatment. A suspected snakebite victim should be muzzled at the outset. Signs of muscular weakness and ataxia may be seen within a few minutes or may be delayed for a few hours. Tetanic spasms may be observed in some cases. Vocalization becomes impaired early in the course of the paralysis, as does swallowing.

Other organ systems may be involved. Vomition is commonly observed with a tiger snake bite, but rarely with a bite from a red-bellied black snake. Constipation and retention of urine are observed after a few hours. Hemorrhage is prominent in bites of the copperhead and the red-bellied black snake. Signs of hypovolemic shock may result from hemorrhage. Anemia may be present, caused by hemorrhage or hemolysis in dogs that live for more than 24 h and

TABLE 2
Summary of Venoms of Selected Australian Snakes

Common name	Scientific name	Yield of venom (mg, dry)	Toxicity in mice (mg/kg)	Action neurotoxic	Action hemolytic cytotoxic	Action coagulant
Common brown	*Pseudonaja textilis*	2	0.02	+3	+1	+3
Copperhead	*Austrelaps superba*	25	0.5	+2	+2	0
Tiger snake	*Notechis scutatus*	35	0.26	+3	+1	0
Death adder	*Acanthophis antarcticus*	40	0.4	+3	+1	0
Red-bellied black snake	*Pseudechis porphyriacus*	40	1.8	+1	+3	+2
Taipan	*Oxyuranus scutellatus*	120	0.12	+3	+1	+1
King brown	*Pseudechis australis*	180	2.4	+1	+1	+1

TABLE 3
Relative Lethal Dose of Tiger Snake, *Notechis scutatus*, Venom When Administered Subcutaneously in Various Species

Species	Dose (mg/kg)
Horse	0.005
Sheep	0.01
Goat	0.018
Monkey	0.02
Guinea pig	0.02
Rabbit	0.045
Cat	0.1
Mouse	0.25
Rat	0.4

Modified from Lewis, P. F., *Refresher Course for Veterinarian, Proc. No. 36, Post-graduate Committee in Veterinary Science*, Hungerford, T. G., Ed., 1978, 287.

these patients are likely to become icteric and hemoglobinuric. Pupillary dilatation is a frequently observed clinical sign in Australian snakebite.

Most lethal bites result in death within 24 h. Even if the dog survives this period, residual effects of hemolysis, renal toxicity, and hemorrhage may result in death 6 to 8 d later.[1,25]

B. CAT[14-16,23,26]

Cats are not envenomated as frequently as dogs, partly because cats are less likely to attack a snake. Cats also appear to be more resistant than dogs to the effects of some Australian snake venoms. The prognosis for a snake-bitten cat is more favorable than for a dog. Progressive flaccid paralysis, pupillary dilatation, coma, and respiratory paralysis are signs of lethal envenomation in the cat.

C. HORSE[1,7,18,25,27,28]

Horses are the most susceptible to tiger snake venom of any species studied (Table 3).[28] A lethal bite in a horse requires a total of only 2.5 mg of venom and a tiger snake may inject 35 mg in a single bite, which is enough venom to kill 14 horses. Kellaway studied the effects of tiger snake venom on the horse and reported sweating, muscular twitching, and paralysis, beginning with the hindquarters and progressing craniad over a period of 6 to 20 h. Other signs included profuse sweating, hemoglobinuria, excessive salivation, pupillary dilatation, and dyspnea, culminating in respiratory paralysis and death.[29] Fitzgerald reported similar signs in horses.[27]

Common brown snake envenomation in the horse has also been studied experimentally by Kellaway.[13] Signs commenced in 3 h, with depression and hemorrhage from the nostrils, followed by progressive paralysis, dyspnea, and death. Pascoe[7] reported on 16 snake bites from one farm during a 6-year period.

Most of the bites were in foals less than 50 d old and were attributed to the common brown snake. Clinical signs included inability to swallow (milk flowing from the nose when attempting to nurse), engorged mammary gland of the mare, stiff-legged (goose-stepping) gait, drooping of the upper eyelid and lower lip, general weakness, dyspnea, recumbency, respiratory paralysis, and death in as short a time as 6 h to as long as 6 d. Just prior to recumbency, muscle twitching, tachycardia, and labored breathing became pronounced, but these signs subsided once the foal was recumbent.

Adult horses in the series of envenomations in Queensland exhibited excessive salivation, inability to swallow, foul-smelling breath from food packed in the mouth, dyspnea, injected mucous membranes, profuse sweating, staggery gait, recumbency, and death. Some horses had ptosis and lower lip paresis.[7]

Subsequent to Pascoe's report on brown snakebite on a Queensland stud farm, he diagnosed botulism in horses on the same farm. Clinical signs were quite similar to brown snakebite. Pascoe was successful in treating most of the suspected snakebite victims with antivenin, but a few cases did not respond to antivenin therapy. Retrospective evaluation suggests that those animals unresponsive to antivenin may have actually been suffering from botulism.*

The red-bellied black snake is unlikely to kill an adult horse, but may produce local edema, hemorrhage at the bite site, and hematuria.[25]

D. OTHER SPECIES

Sheep and goats were used as experimental animals by Kellaway.[10-13] Tiger snake venom caused muscular twitching, paralysis of the tongue and general body musculature, dribbling of saliva as a result of inability to swallow, regurgitation of rumen contents through the mouth and nostrils, and intravascular clotting.

Cattle die from the effects of snake venom, but precise documentation of the syndrome is lacking.[25,30] Swine are not immune to the effects of Australian snake venoms, but their tough, thick hide may preclude penetration by any but the largest snakes and even when the hide is penetrated, the subcutaneous fat layer may inhibit absorption of the venom so that clinical signs rarely occur.[25]

There is poor documentation of snakebite in wild animals. One report mentioned an unspecified wallaby as an experimental subject.[14] Wild animals are probably envenomated, but nothing is known of the resistance of Australian wild mammals to snake venoms.

V. DIAGNOSIS

Clinical signs are seldom precise because of marked variation in syndromes, even from bites of the same species of snake. Clinicians usually must base a diagnosis on the species present in the practice area, predominant clinical signs exhibited, and previous experience. The development of hemorrhage, hemolysis, and coagulant disorders tend to implicate black snakes and the copperhead.

* Pascoe, R., personal communication, February, 1992.

Differential diagnosis must include envenomation by the scrub or bush tick *(Ixodes holocyclus),* insecticide poisoning, botulism, acute infectious diseases such as canine hepatitis or leptospirosis, and any disorder producing hematuria, myoglobinuria, or hemoglobinuria.

Hemorrhages are the most consistent lesions found at necropsy.

Much effort has been expended to develop immunologic tests for identification of snake venoms so that specific monovalent antivenins may be used. Results of radioimmunoassay (RIA) were precise, but too slow, and required specialized laboratory equipment and reagents. Recently, the Commonwealth Serum Laboratories have developed field kits for use in clinical situations. Venom may be identified from swabs of the bite site or from blood samples.

VI. TREATMENT

If a lethal dose of venom has been injected, supportive and symptomatic treatment will be of no avail without timely, adequate administration of antivenin. See Chapter 13 for a list of the antivenins available in Australia and the methods and precautions for administration. Such therapy is costly but crucial. The key is to begin antivenin administration as soon possible after any signs of envenomation appear. This may result in overtreatment of mildly envenomated patients, but delay may allow neurotoxins to produce irreversible changes that may threaten the life of the animal despite antivenin therapy.

Monovalent antivenenes are produced for the death adder, eastern brown snake, taipan, and king brown snake.[8] A combined tiger snake/sea snake antivenene is available and a polyvalent antivenene has been produced but is quite expensive. When the identity of the snake is known, it is best to use a monovalent antivenene. When the identity is unknown, antivenins are administered according to geographical regions. In Tasmania, tiger snake/sea snake antivenene (6000 units) is used. In Victoria, both tiger snake (3000 units) and eastern brown snake antivenene (1000 units) are used. In other states, the polyvalent antivenene is recommended because it includes venoms from more species of snakes, any of which may be involved. Monovalent antivenenes may not provide cross-protection.

REFERENCES

1. **Lewis, P. F.**, Snakebite in animals in Australia, in *Refresher Course for Veterinarian, Proc. No. 36*, Hungerford, T. G., Ed., *Post-graduate Committee in Veterinary Science*, Sydney, Australia, 1978, 287.
2. **Wilson, S. K. and Knowles, D. G.**, *Australia's Reptiles*, Collins, Sydney, Australia, 1988.
3. **Mirtschin, P. and Davis, R.**, *Dangerous Snakes of Australia*, Revised ed., Rigby Publishers, Dee Why West, New South Wales, Australia, 1983.

4. **Barme, M.,** Venomous sea snakes (Hydrophiidae), in *Venomous Animals and Their Venoms, Vol. 1. Venomous Vertebrates,* Bucherl, W., Buckley, E. E., and Deulofeu, V., Eds., Academic Press, Orlando, FL, 1968, 286.
5. **Halstead, B. W.,** *Poisonous and Venomous Marine Animals of the World,* Revised ed., Darwin Press, Princeton, NJ, 1978.
6. **Tu, A. T.,** Sea snakes and their venoms, in *Handbook of Natural Toxins, Vol. 3. Marine Toxins and Venoms,* Tu, A. T., Ed., Marcel Dekker, New York,
7. **Pascoe, R.,** Brown snake bite in horses in south-eastern Queensland, *J. S. Afr. Vet. Assoc.,* 46, 129, 1975.
8. **Chippaux, J.-P. and Goyfoon, M.,** Production and use of snake antivenin, in *Handbook of Natural Toxins, Vol. 5. Reptile Venoms and Toxins,* Tu, A. T., Ed., Marcel Dekker, New York, 1991. 529.
9. **Sutherland, S. K.,** *Australian Animal Toxins,* Oxford University Press, Melbourne, Australia, 1983.
10. **Kellaway, C. H.,** A preliminary note on the venom of the Australian copper-head *Denisonia superba:* its toxic effect on the common laboratory animals, *Med. J. Austr.,* 1, 358, 1929.
11. **Kellaway, C. H.,** Observations on the certainly lethal dose of the venom of the death adder *Acanthophis antaracticus* for the common laboratory animals, *Med. J. Austr.,* 1, 764, 1929.
12. **Kellaway, C. H.,** Observations on the certainly lethal dose of the venom of the black snake *Pseudechis porphyriacus* for the common laboratory animals, *Med. J. Austr.,* 2, 33, 1930.
13. **Kellaway, C. H.,** Observations on the certainly lethal dose of the venom of the common brown snake *Demansia textilis* for the common laboratory animals, *Med. J. Austr.,* 2, 747, 1931.
14. **Crawford, A. M. and Mills, J. N.,** Three Western Australian snake venoms on blood coagulation of the dog, cat, horse and wallaby, *Austr. Vet. J.,* 62, 185, 1985.
15. **Holloway, S. A. and Parry, B. W.,** Observations on blood coagulation after snakebite in dogs and cats, *Austr. Vet. J.,* 66, 364, 1989.
16. **Shea, G. M.,** Three Western Australian snake venoms on blood coagulation of the dog, cat, horse and wallaby — a criticism of the paper by Crawford et al., *Austr. Vet. J.,* 63, 352, 1986.
17. **Lewis, P. F.,** Tiger Snake envenomation in the Dog, Ph.D. thesis, University of Melbourne, Melbourne, 1984.
18. **Hungerford, T. G.,** Snakebite, in *Diseases of Livestock,* 8th ed., Hungerford, T. G., Ed., McGraw-Hill, Auckland, 1975, 1079.
19. **Dickinson, D. A.,** Snake bite in a dog, *Austr. Vet. J.,* 29, 170, 1953.
20. **Fethers, G.,** Tiger snake *Notechis scutatus* bite in a dog, successfully treated with antivenene, *Austr. Vet. J.,* 19, 45, 1943.
21. **Gordon, R. N.,** The symptoms, diagnosis and treatment of snakebite in the dog, *Austr. Vet. J.,* 34, 216, 1958.
22. **Hill, F. W. G. and Campbell, T.,** Snake bite in dogs, *Austr. Vet. J.,* 55, 82, 1979.
23. **Hirschhorn, H. L.,** Snakebite in Australia and New Guinea, in *Current Veterinary Therapy, VII. Small Animal Practice,* Kirk, R. W., Ed., W. B. Saunders, Philadelphia, 1980, 179.
24. **Keep, J. M.,** Antivenene in a case of snake bite in the dog, *Austr. Vet. J.,* 33, 95, 1957.
25. **Seawright, H. L.,** Snakebite, in *Diseases of Livestock,* 8th ed., Hungerford, T. G., Ed., McGraw-Hill, Auckland, 1975, 1079.
26. **Hill, F. W. G. and Campbell, T.,** Snake bite in cats, *Austr. Vet. J.,* 54, 437, 1978.
27. **Fitzgerald, W. E.,** Snakebite in the horse, *Austr. Vet. J.,* 51, 37, 1975.
28. **Prescott, C. W.,** The snake venom intoxications — 1, *Austr. Vet. Practitioner,* 14, 53, 1984.
29. **Kellaway, C. H.,** The venom of *Notechis scutatus, Med. J. Austr.,* 1, 348, 1929.
30. **Couttie, P. M.,** Treatment of snakebite in calves, *Austr. Vet. J.,* 45, 384, 1969.

Chapter 18

SNAKES AND SNAKEBITE IN EUROPE

I. BIOLOGY AND IDENTIFICATION

European countries, particularly those in the northern temperate regions, have a sparse venomous snake fauna with species only in the family *Viperidae* represented (Table 1). The prevalence of venomous snakebite in the human population is low and fatality rare.[1,2] Venomous snakebite in animals is of minimal clinical significance, but large and more aggressive individuals of the European viper *(Vipera berus)*, asp viper *(Vipera aspis)*, long-nosed viper *(Vipera ammodytes)*, and Orsini's viper *(Vipera ursinii)* may inflict bites that require veterinary care.

The European viper is the only venomous snake found in northern Europe, Scandinavia, England, and Scotland.[3-5] It is distributed south to northern Spain and east throughout central Europe to northern Asia, including North Korea (Figure 1). The European viper is a small snake averaging 0.48 to 0.61 m (19 to 24 in.) to a maximum of 0.86 m (34 in.) in length. The body color is pale gray, olive or yellow to russet or brown, with a dark brown or black zigzag line or series of dots extending along the back. The head is oval in outline rather than the more typical triangular shape characteristic of vipers.[3]

The general distribution of the asp viper is illustrated in Figure 2. The asp viper has a more triangular head than the European viper, but is of similar size and body color.[3] The asp viper is more phlegmatic than the European viper and is less likely to be involved in an animal bite unless attacked by a dog.

The long-nosed viper is distributed throughout southeastern Europe and Asia Minor.[3] It is of similar size to the other European vipers. The head is distinctive, with an elongated snout that ends in a pronounced upturned appendage.[3] Its basic body color ranges from ash-gray, to yellow, pale orange, coppery, or brownish with a prominent zigzag band down the back.

Orsini's viper is restricted to southeastern Europe (Figure 2). Its size and shape are similar to other vipers in the area. It is rarely involved in human fatalities and is unlikely to inflict bites of clinical significance in animals.

II. CONDITIONS OF POISONING

The European climate varies from subarctic in the north to Mediterranean in the south. Snakes adapt to the variable climates. The European viper hibernates in the north; thus, snakebite is a problem only during the warm spring and summer months. In the south, the species may be active all year. When the weather is warm to hot, the snake may be nocturnal, but in cool weather be diurnal.[3] The European viper is generally timid, but may strike rapidly and repeatedly if agitated.[3]

TABLE 1
Selected Venomous Snakes — Europe

Phylum — *Chordata*
Class — *Reptilia*
Order — *Squamata*
Suborder — *Ophidia (Serpentes)* — snakes
  Family — *Elapidae*
  Family — *Viperidae* — vipers, adders
    Vipera berus — European adder
    Vipera ammodytes — long-nosed viper
    Vipera ursinii — Orsini's viper

The long-nosed viper is primarily nocturnal, but sedentary and retiring unless provoked.[3]

III. VENOM

The toxicity of the venom of these vipers varies considerably throughout the region, but generally bites result in local or regional cytotoxic and necrotoxic effects.

IV. SIGNS OF ENVENOMATION

A. HUMAN

Pain is usually experienced at the time of the bite, but it may be delayed for as long as 15 min when venom has been injected. Edema appears at the bite site within minutes and may spread regionally for the next 48 to 72 h. The skin may become discolored and vesicles and subcutaneous hematomas may be formed. Localized necrosis may occur, but rarely necessitates amputation. Secondary infection of the devitalized tissue is common.[1]

Systemic signs may include nausea, vomition, hematemesis, colic, melena, and diarrhea. Neurotoxic, cardiotoxic, and nephrotoxic signs are usually mild or absent.[1]

B. ANIMAL[6-11]

The clinical signs of European adder envenomation in the dog include local, painful swelling, vomiting, dehydration, depression, and death in 5 to 7 d in untreated dogs.[7,8] Muzzle bites in horses result in swelling and potential obstruction of the nares. Gonzalez[8] reported on 375 cases of viper bites in domestic animals in Spain (281 dogs, 15 cats, 45 sheep, 3 goats, 7 cows, 2 horses, 1 mule, 2 donkeys, and 17 fowl).[8] He reported that mortality was high in dogs and cats because of their size, and that fatal bites occurred on the mouth, tongue, and face.[8] Clinical signs included edema and hemorrhage at the bite site, necrosis, inability to use a limb for a long period of time, and coagulation disorders.

FIGURE 1. Distribution of European viper, *Vipera berus*. (Modified from Klemmer, K., *Venomous Animals and Their Venoms, Vol. 1. Venomous Vertebrates*, Bucherl, W., Buckley, E. E., and Deulofeu, V., Eds., Academic Press, Orlando, FL, 1968. With permission.)

Kängstrom[9] reported on 170 dogs and 54 cats bitten by the European viper. Bites were most common on the face, lips, nose, and forefeet. The major sign was severe local hemorrhagic edema. Animals that died usually became anemic, uremic, thrombocytopenic, hypoproteinemic, and hypoalbuminemic. The mortality rate for dogs and cats was 3.5%.[9]

V. DIAGNOSIS

Snakebite in animals in Europe is so infrequent that clinicians may have little experience with the clinical manifestations. Rarely will a history of snakebite be forthcoming from an owner. Diagnosis is based on the syndrome of local edema, hemorrhage, vesiculation, and, possibly, necrosis. Clinical pathology values are not likely to be of assistance in diagnosing an animal bite.

At necropsy, lesions resulting from envenomation by the European viper included hemorrhages, disseminated intravascular coagulations, hepatic necrosis and renal degeneration, and infarction.[9] Lesions observed by Bratberg and Flesja[7] at necropsy included hemorrhagic edema at the bite site, pulmonary congestion, generalized hemolysis, icterus, myocardial degeneration, and petechial hemorrhages in the kidney cortex.[7] Similar lesions were observed in dogs dying from envenomation by the long-nosed viper.[7]

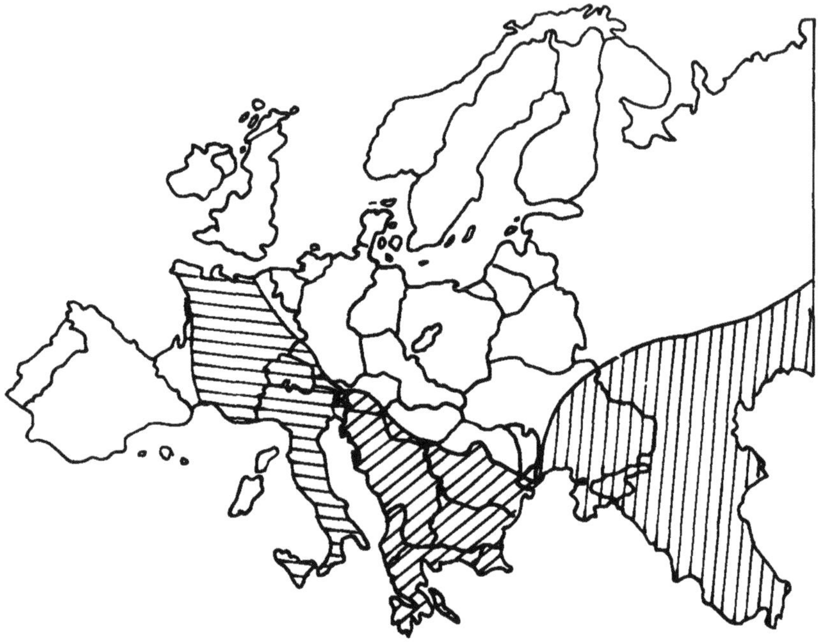

FIGURE 2. Distribution of: ▤European asp, *Vipera aspis.* ▥Long-nosed viper, *Vipera ammodytes.* ▨Orsini's viper, *Vipera ursinii.* (Modified from Klemmer, K., *Venomous Animals and Their Venoms, Vol. 1. Venomous Vertebrates,* Bucherl, W., Buckley, E. E., and Deulofeu, V., Eds., Academic Press, Orlando, FL, 1968. With permission.)

VI. TREATMENT

A. HUMAN

Administration of specific antivenin is the optimal therapy, accompanied by careful monitoring of the patient for systemic signs. The bite site should be treated as for other viper bites (Chapter 14, page 121).

B. ANIMAL

Antivenin therapy is indicated in animals as well, but will rarely be employed. The local reaction is managed as for other viper or crotalid bites.

REFERENCES

1. **Gonzalez, D.,** Snakebite problems in Europe, in *Handbook of Natural Toxins, Vol. 5. Reptile Venoms and Toxins,* Tu, A. T., Ed., Marcel Dekker, New York, 1991, 687.
2. **Swaroop, S. and Grab, B.,** Snakebite mortality in the world, *Bull. World Health Org.,* 10, 35, 1954.
3. **Moore, G. M., Ed.,** *Poisonous Snakes of the World,* NAVMED P-5099, Revised 1965, Dept. of the Navy, Bureau Medicine and Surgery, U.S. Government Printing Office, Washington, D.C., 1965.
4. **Gitter, S. and de Vries, A.,** Symptomatology, pathology and treatment of bites by Near Eastern, European and North African snakes, in *Venomous Animals and Their Venoms, Vol. 1. Venomous Vertebrates,* Bucherl, W., Buckley, E. E., and Deulofeu, V., Eds., Academic Press, Orlando, FL, 1968, 359.
5. **Klemmer, K.,** Classification and distribution of European, North African and North and West Asiatic venomous snakes, in *Venomous Animals and Their Venoms, Vol. 1. Venomous Vertebrates,* Bucherl, W., Buckley, E. E., and Deulofeu, V., Eds., Academic Press, Orlando, FL, 1968, 309.
6. **Gonzalez, D.,** The epidemiological aspects of viper bites in Europe, *Toxicon,* 28, 149, 1990.
7. **Bratberg, B. and Flesja, K.,** Intoksikasjon som folge av hoggormbitt (Intoxication in the dog by the European adder), *Nord. Vet.-Med.,* 25, 185, 1973.
8. **Gonzalez, D.,** Snake bites in domestic animals, *Toxicon,* 28, 149, 1990.
9. **Kängstrom, L.-E.,** Huggormsbett hos hund och katt (Snakebite in the dog and cat), *Svensk Veterinartidning,* 41, 38, 1989. (Swedish)
10. **Bartik, M. and Piskač, A.,** *Veterinary Toxicology,* Elsevier, Amsterdam, 1981, 297.
11. **Humphreys, D. J.,** *Veterinary Toxicology,* 3rd ed., Bailliere Tindall, London, 1988, 313.

Chapter 19

SNAKEBITE IN AFRICA

I. VENOMOUS SNAKES

Hundreds of different habitats on the African continent vary from the Saharan desert regions of North Africa to the tropical forests of central and west Africa. As might be expected, the snake fauna is highly complex; 137 species are reported to be venomous, but fewer than 25 species in 4 snake families *(Viperidae, Elapidae, Colubridae,* and *Hydrophiidae)* are likely to cause human mortality.[1-4]

The prevalence of snakebite in animals is unknown. Species likely to cause the majority of animal bites in Africa are listed in Table 1. The selection is based on size, potency of venom, distribution, aggressiveness, and habitat selection.

Sea snakes are found in coastal African waters, but are not involved in animal envenomation. General information on sea snakes may be found in Chapter 17. Rear-fanged colubrid snakes may envenomate humans, but there are no reports of animal bites from this group of snakes.

A. DESCRIPTION OF SELECTED SPECIES

1. Family *Elapidae*

The rinkals cobra, *Hemachatus haemachatus,* is confined to the southern tip of Africa. It is an accurate spitter and is considered to be one of the more dangerous venomous snakes within its range. Adults average 1.0 to 1.2 m in length, with a maximum of 1.4 m. The basic body color is dark brown to black with irregular crossbands of lighter brown.[5] The fangs are short, but muscular apparatus around the venom gland provides compression for driving the venom accurately for as far as 2 m.

A number of species of the genus *Naja* inhabit Africa. The black spitting cobra, *Naja nigricollis,* is distributed throughout much of subSaharan Africa except for South Africa (Figure 1). Adults have variable body color, from pinkish-tan to uniform black and have an average length of 1.5 to 1.8 m up to a maximum of 2.25 m.[5]

The Egyptian cobra, *Naja haje,* has the broadest distribution of any venomous species on the continent (Figure 2). Egyptian cobras vary in color from light to dark brown or black and some specimens have prominent banding.[2] Egyptian cobras average 1.2 to 1.8 m in length, with a maximum of 2.4 m (8 ft). The hood is prominent in its defensive posture.

Although there are a number of species of mambas, the most notorious African elapid is the black mamba, *Dendroaspis polylepis,* (Figure 2). The name "black mamba" is derived from the black oral mucosa, not the body color which is olive brown to dark gray. Adult black mambas average 2.7 to 3.0 m in length up to a maximum of 4.3 m.[5] The black mamba is considered to be thc

TABLE 1
Selected Venomous Snakes — Africa

Phylum — *Chordata*
Class — *Reptilia*
Order — *Squamata*
Suborder — *Ophidia (Serpentes)* — snakes
  Family — *Elapidae* — cobras, spitting cobras, mamba
    Hemachatus haemachatus — rinkals
    Naja haje — Egyptian or brown cobra
    Naja nigricollis — spitting cobra
    Dendroaspis polylepis — black mamba
  Family — *Viperidae* — vipers, adders, asps
    Bitis arietans — African puff adder
    B. gabonica — Gaboon viper (adder)
    B. nasicornis — rhinoceros viper
    Cerastes vipera — Cleopatra's asp
    Echis carinatus — saw-scaled viper

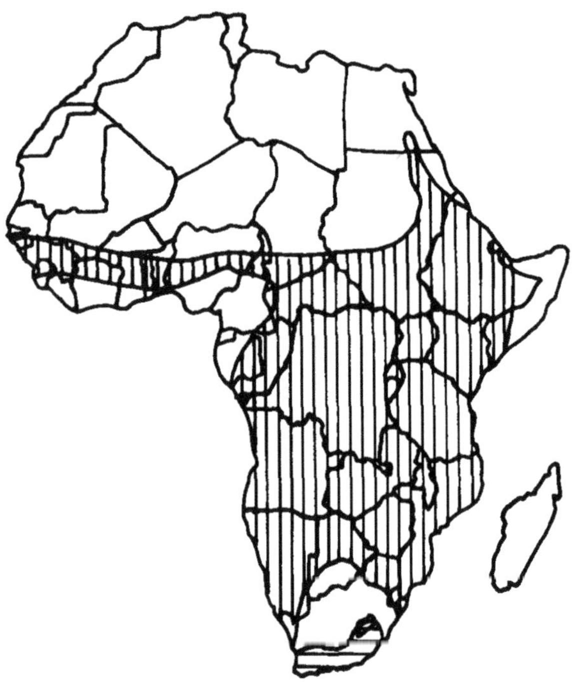

FIGURE 1. Distribution of: ▤Rinkals, *Hemachatus haemachatus*. ▥ Spitting cobra, *Naja nigricollis*, (Modified from Broadley, D. G., *Venomous Animals and Their Venoms, Vol. 1. Venomous Vertebrates*, Bucherl, W., Buckley, E. E., and Deulofeu, V., Eds., Academic Press, Orlando, FL, 1968. With permission.)

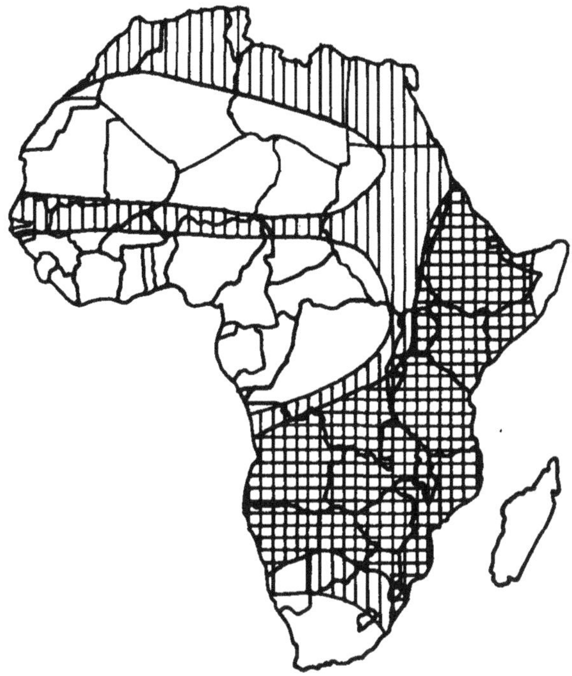

FIGURE 2. Approximate distribution of: ▤Black mamba, *Dendroaspis polylepis*. ▥Egyptian cobra, *Naja haje*. (Modified from Broadley, D. G., *Venomous Animals and Their Venoms, Vol. 1. Venomous Vertebrates*, Bucherl, W., Buckley, E. E., and Deulofeu, V., Eds., Academic Press, Orlando, FL, 1968. With permission.)

fastest snake, reaching speeds of 20 kph (12 mph). The black mamba's legendary sinister reputation for aggression is probably the result of the lethality of its bite. In fact, the black mamba is a shy snake that will move away if possible.[2] If cornered, it assumes a defensive posture with the head raised well off the ground, mouth agape showing the black lining, and the tongue flicking from side to side.[5] A thin hood is evident. If the snake is not immediately left alone, it may dash for freedom past the enemy, striking as it passes.[2] It is capable of striking a distance of 40% of its body length. The black mamba is one of the most dangerous snakes now living, secreting sufficient venom to kill 5 to 10 humans. Although the snake does not bite and hold on, it may rapidly inflict repeated bites if highly agitated. Few humans survive without immediate antivenin treatment.

The author lived for a time in a grass hut in the Okavango Swamp of Botswana. The previous occupant had died from the bite of a black mamba in the hut. Ironically, although antivenin was in the camp, no syringe or needle was available to administer it.

FIGURE 3. African puff adder, *Bitis arietans*.

2. Family *Viperidae*

The African puff adder, *Bitis arietans*, is a nocturnal, slow-moving, heavy-bodied viper with a lanceolate head. Its average length is 0.9 to 1.2 m ranging in length to a maximum of 1.5 m. Body color is alternating dark and light chevron-shaped markings[5] (Figure 3). The puff adder is broadly distributed in Africa (Figure 4), and because of its nocturnal behavior, large size, and potent venom, probably kills more humans than any other African snake.[5]

The Gaboon viper, *Bitis gabonica*, (Figure 5), is less broadly distributed (cf. Figures 4 and 6) than the puff adder, but it too is a heavy-bodied, large snake with an average length of 0.9 to 1.2 m and a maximum of 2.0 m, making it the largest of the vipers.[5] Body color is distinctive, a blotchy pattern of browns and tans, with a tan crown of the head and a median narrow brown stripe along the back. The fangs on this snake are 5 cm (2 in.) long.[5]

The rhinoceros viper or river jack, *Bitis nasicornis* (Figure 7), is slightly smaller than the previous two vipers, but is still an impressive snake.

The saw-scaled viper, *Echis carinatus*, ranges from Sri Lanka and southern India to western Asia, the Middle East, and North Africa southward to tropical central Africa (Figure 6). Although a small snake, less than 0.9 m long, its broad distribution and highly toxic venom make it dangerous. The name derives from its behavior when threatened, which is to inflate the lung and air sac to full capacity, expanding the body, then rubbing the everted lateral scales together to produce a saw-like hissing sound.[5]

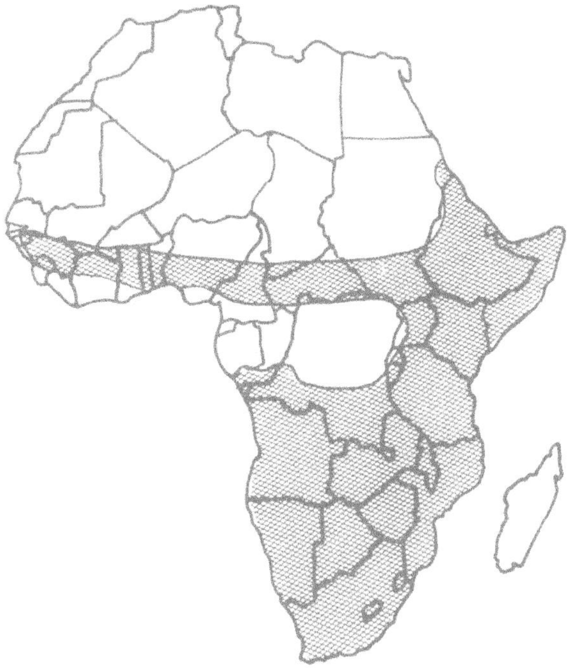

FIGURE 4. Distribution of African puff adder, ***Bitis arietans***. (Modified from Broadley, D. G., *Venomous Animals and Their Venoms, Vol. 1. Venomous Vertebrates*, Bucherl, W., Buckley, E. E., and Deulofeu, V., Eds., Academic Press, Orlando, FL, 1968. With permission.)

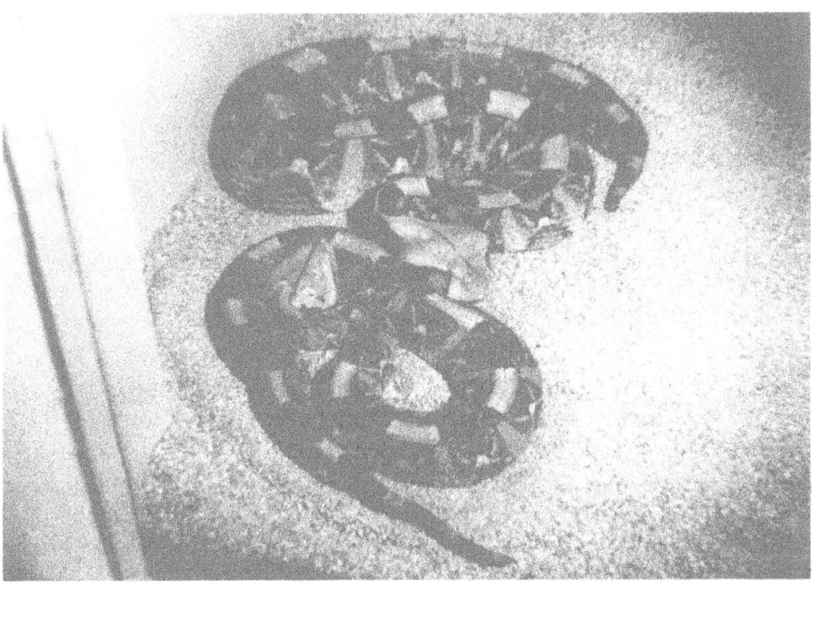

FIGURE 5. Gaboon viper, *Bitis gabonica*.

FIGURE 6. Distribution of: ▥ Saw-scaled viper, *Echis carinatus*. ▤ Gaboon viper, *Bitis gabonica*.

II. CONDITIONS OF ENVENOMATION

A. HUMAN

Some venomous species may be found around human habitations and most of them are nocturnal. Humans moving about in early evening or at night are at risk as they may step near or on the snake. Pastoralists may come in contact with snakes in the fields. Hikers or field biologists may likewise encounter snakes.

B. ANIMAL

Statistical data on the extent of animal envenomation is minimal, but livestock and companion animals are essentially at the same risk as humans. Mason reports that dogs are usually bitten on the head, while horses and cattle are bitten on the head and lower legs.[6] Spitting cobras use this behavior in defense. Most of the spitters dwell in open savannah grassland. It is thought that spitting may have evolved as a means of dissuading plains ungulates from treading upon them.

FIGURE 7. Rhinoceros viper, *Bitis nasicornis*.

III. VENOM[5-9]

Elapid snake venom is primarily neurotoxic. The seriousness of envenomation is determined primarily by the volume of venom injected. Viper venom is more complex than elapid venom, but is primarily cytotoxic, proteolytic, necrotoxic, and anticoagulant.[2] Blood loss may be significant and must be treated if therapy is to be successful.

IV. SIGNS OF ENVENOMATION

A. HUMAN[2,10-12]

1. Elapid Bites

Primary signs are labored breathing and excessive salivation caused by paralysis of the muscles of respiration and deglutition. Other signs of neural origin are vertigo, restlessness, convulsions, drowsiness, ptosis, strabismus, and slurred speech. Death may occur within a few minutes of the bite or be delayed as long as 12 h in untreated victims.[2] If the patient survives for 24 h, few residual effects are noted.

The bite of a black mamba is greatly feared because of the rapidity of the development of signs and sure progress to death if antivenin is not administered within an hour.[2] The venom of the black mamba is not markedly more toxic than other elapid venoms (Table 2), but the volume injected is greater.

TABLE 2
Relative Toxicity of Selected African Snake Venoms

Snake	LD_{50} (mg/kg mouse, i.v.)
Naja haje Egyptian cobra	0.42
Naja nigricollis Spitting cobra	0.72
Hemachatus haemachatus Rinkals, spitting cobra	1.70
Dendroaspis polylepis Black mamba	0.25
Bitis arietans African puff adder	0.42–2.0
Bitis gabonica Gaboon viper	4.9
Bitis nasicornis Rhinoceros viper, riverjack	8.6
Echis carinatus Saw-scaled viper	0.4–0.9

Modified from Khole, V., *Handbook of Natural Toxins, Vol. 5. Reptile Venoms and Toxins*, Tu, A. T., Ed., Marcel Dekker, New York, 1991. With permission.

Spitting is a defensive maneuver, blinding the victim and allowing the snake to escape. The reaction of the conjunctival sac to the venom is production of an immediate, intense keratitis and conjunctivitis with subsequent lacrimation, exudation, pain, and photophobia.[2] The intense irritation usually abates within 2 or 3 d, with no after effects and no evidence of venom absorption.[2] If a spitting cobra is stepped upon, it can and will bite, the venom producing similar signs to other elapid snakebites.

2. Viperid Bites

Principal signs resulting from viper bites are mild to severe tissue swelling, shock, hemorrhage, tissue necrosis, and secondary infection. Blood loss may be severe.

B. ANIMAL

Christensen[7] reported on 98 animal envenomations (dogs 60%, oxen 30%, horses 3%, sheep 2%, and cats 2%) in South Africa. The more important snakes involved were the puff adder, rinkals, cape cobra, black mamba, and other adders.

In the veterinary literature, reference to snakebite in animals is sparse.[6,7] Yet, in discussion with veterinarians from African countries, the response to questions about snakebite is, "Yes, snakebite occurs." The clinical signs in animals would probably mimic, to a great extent, those seen in humans.

No published information exists on the prevalence of snakebite in wild animals, but a fortuitous filming of African lions in Botswana provided a vivid record of the effects of a bite by an Egyptian cobra. Two National Geographic photographers, on assignment to record the interaction of lions and hyenas, chanced to film a lioness as she moved her three small cubs, only a few weeks old, from one den to another. Unbeknownst to the lioness, the new site was near the lair of the cobra. When all the cubs had been moved, the inevitable encounter occurred and the cobra confronted the lioness and struck her. The photographers also witnessed the cobra fatally biting, but not attempting to consume, the cubs.

The effects of the neurotoxic venom were severe. Within minutes, the lioness exhibited ataxia and drooling of saliva. Ultimately, she became blind and tetraplegic. Though separated from her pride and at the total mercy of other predators, the lioness was not killed. She remained recumbent for days, slowly regaining her sight and equilibrium. She managed to rejoin her pride, remaining under their protection until she was fit to participate in hunts and kills.

V. DIAGNOSIS

A. HUMAN

It is not difficult to differentiate between the neurotoxic signs of elapid envenomation and the cytotoxic signs produced by a viperid bite. However, the species identity of the snake is rarely known because the victims are frequently children or farm laborers with sparse knowledge of snake identification and the bites usually occur at night. Treatment usually proceeds on a symptomatic basis.

B. ANIMAL

There are no criteria for diagnosis of snakebite in animals other than similarity to signs seen in humans.

VI. TREATMENT

A. HUMAN

1. Elapid Bites

First aid for a known elapid bite is prompt application of a broad tourniquet (2 to 5 cm wide) and transport of the victim to a medical facility. The tourniquet should be released every 15 min. If the medical facility is too far away, and if a polyvalent antivenin is available, administer 10 ml intramuscularly. Be prepared to administer cardiopulmonary respiration and keep the head positioned to prevent aspiration of saliva.[2]

The key to successful treatment of an elapid bite is prompt intravenous administration of large volumes of antivenin: specific monovalent if the snake is known, or polyvalent antivenin if it is not. Respiratory failure is the proximate cause of death, so support of respiration is crucial. Suction should be provided to remove saliva from the mouth to prevent inhalation and drowning of the victim.[2]

Recommended treatment of conjunctivitis caused by spitting cobra spray is lavage of the conjunctival sac with large volumes of normal saline plus administration of analgesics.[2]

2. Viperid Bites

The physician dealing with a viper bite must counter the effects of swelling caused either by the cytotoxic action of the venom or extravasation of blood. Shock and anemia caused by blood loss, tissue necrosis, and secondary infection are complications.[2] Visser and Chapman's review[2] of cases treated in South Africa is a primary source of information on the subject, but it must be recognized that the work was published in 1978 and more recent experiences in the treatment of snakebite may modify the information presented in that text. For instance, Visser felt that antivenin had little or no effect on the cytotoxic action of viperid venom. Recent studies on the treatment of crotalid bites (crotalid venom is also cytotoxic) have shown that antivenin, administered promptly in appropriate volumes, is effective. However, antivenin cannot correct cytotoxic effects that have already developed. If tissue necrosis, thrombosis, or extravasation of blood has already taken place, antivenin will not be helpful.

Hemorrhage may be visible on the surface or occur internally within the muscle or deeper tissues. Blood transfusions are an important component of therapy for African viperid bites. Blood loss may be difficult to estimate; a hemogram and total protein are indicated.

An estimate of superficial blood loss can be made on the basis that a bruise the size of a human fist equals 500 ml of blood.[2] Or, "...the first slight swelling, large enough to be visualized in an area of the body, requires an increase of 10% of the original volume of that part."[2] This rule of thumb could be applied to animal bites as well.

Tissue necrosis may result from cytotoxic effects, from ischemia caused by thrombosis of vessels, or from pressure and occlusion of the vessels as a result of tissue swelling. Debridement and even amputation may be necessary to avoid systemic absorption of toxins from necrotic tissue. Administration of broad-spectrum antibiotics is indicated to prevent secondary infection and sepsis.

B. ANIMAL

The use of adequate amounts of antivenin as soon as possible is advocated.[6] Animal bites should probably be treated similarly to human snakebite.

REFERENCES

1. **Broadley, D. G.,** The venomous snakes of Central and South Africa, in *Venomous Animals and Their Venoms, Vol. 1. Venomous Vertebrates,* Bucherl, W., Buckley, E. E., and Deulofeu, V., Eds., Academic Press, Orlando, FL, 1968, 403.
2. **Visser, J. and Chapman, D. S.,** *Snakes and Snakebite,* Purnell, Cape Town, South Africa, 1978.
3. **Caras, R.,** *Venomous Animals of the World,* Prentice-Hall, Englewood Cliffs, NJ, 1974.
4. **Moore, G. M., Ed.,** *Poisonous Snakes of the World,* NAVMED P-5099, Revised 1965, Dept. of the Navy, Bureau Medicine and Surgery, U.S. Government Printing Office, Washington, D.C., 1965.
5. **Russell, F. E. and Saunders, P. P.,** *Animal Toxins,* Pergamon Press, Elmsford, NY, 1967.
6. **Mason, J. H.,** The treatment of snakebite in animals, *J. S. Afr. Vet. Med. Assoc.,* 33, 583, 1962.
7. **Christensen, P. A.,** *South African Snake Venoms and Antivenoms,* Publication of the South African Institute for Medical Research, Johannesburg, 1955, 1-129.
8. **Christensen, P. A.,** The venoms of Central and South African snakes, in *Venomous Animals and Their Venoms, Vol. 1. Venomous Vertebrates,* Bucherl, W., Buckley, E. E., and Deulofeu, V., Eds., Academic Press, Orlando, FL, 1968, 403.
9. **Christensen, P. A. and Anderson, C. G.,** Observations on *Dendroaspis* venoms, in *Animal Toxins,* Russell, F. E. and Saunders, P. R., Eds., Pergamon Press, Oxford, England, 1967, 223.
10. **Russell, F. E.,** Snake venom poisoning, *Vet. Human Toxicol.,* 33, 584, 1991.
11. **Russell, F. E.,** *Snake Venom Poisoning,* Scholium International, Great Neck, NY, 1983.
12. **Chapman, D. S.,** The symptomatology, pathology and treatment of the bites of venomous snakes of Central and South Africa, in *Venomous Animals and Their Venoms, Vol. 1. Venomous Vertebrates,* Bucherl, W., Buckley, E. E., and Deulofeu, V., Eds., Academic Press, Orlando, FL, 1968, 463.
13. **Khole, V.,** Toxicities of snake venoms and their components, in *Handbook of Natural Toxins, Vol. 5. Reptile Venoms and Toxins,* Tu, A. T., Ed., Marcel Dekker, New York, 1991, 405.

Chapter 20

SNAKEBITE IN ASIA, SOUTHEAST ASIA, INDIA, AND THE MIDDLE EAST

I. BIOLOGY OF SELECTED VENOMOUS SNAKES

Venomous snakes inhabiting the countries encompassed by this chapter are numerous and varied. All venomous snake families are represented. Habitats include alpine, montane, tropical, desert, grassland, marsh, littoral, and forest. Species with a limited distribution or insular populations may not be a threat to domestic animals. The author has selected a few species that are considered dangerous to human beings and live in areas likely to be inhabited by domestic livestock and/or companion animals (Table 1).

Information on the prevalence of animal bites in these regions is lacking. However, the assumption is that bites occur and are similar to human bites in clinical presentation and therapy.

A. DESCRIPTION OF SELECTED SPECIES

1. Family *Elapidae*

Some 12 species of kraits, *Bungarus* spp., inhabit Southeast Asia. Large specimens of the banded krait, *Bungarus fasciatus*, may reach a length of 2.13 m (7 ft), but lengths of most kraits range from 1.2 to 1.5 m (4 to 5 ft).[1,2] All kraits are considered dangerous to humans.

There are 13 species of oriental coral snakes, *Calliophis* spp., which inhabit Southeast Asia. The largest specimens may reach 0.9 m (3 ft), but most are small snakes.[2] Like North and South American coral snakes, these species are infrequently encountered because of their secretive habits. Animal bites are unlikely.

The Egyptian cobra, *Naja haje*, may be seen in the Middle East, but the Indian or Asian cobra, *Naja naja naja*, and Oxus cobra, *Naja naja oxiana*, (Figures 1 and 2) are more widespread. Their distribution is illustrated in Figure 3. Other subspecies of *Naja* have limited distribution, but all of them are considered dangerous to humans.

Indian and Oxus cobras have an average adult length of 1.2 to 1.5 m (4 to 5 ft), with a maximum of 2 m (6.5 ft). Body color is brown to black with a light belly and variable patterns. The hood of the Indian cobra is the most pronounced of any of the cobras and dorsal hood marks have the appearance of a pair of spectacles.[1] The hood of the Oxus cobra is narrower and there are no hood marks.[1,2]

The king cobra or hamadryad, *Ophiophagus hannah*, is the largest of the Asian venomous snakes, reaching lengths of 2.1 to 4.0 m (7 to 13 ft) (Figure 3). Body color is a uniform olive-brown to greenish yellow. Legend has it that a king cobra will attack without provocation, but this would be highly unusual

TABLE 1
Selected Venomous Snakes — Asia, Southeast Asia, India, and the Middle East

Phylum — *Chordata*
Class — *Reptilia*
Order — *Squamata*
Suborder — *Ophidia (Serpentes)* — snakes
  Family — *Elapidae* — fixed-front-fanged snakes
    Bungarus multicinctus — krait
    Bungarus spp. — 12+ species
    Calliphis spp. — Asian coral snakes
    Naja haje — Egyptian cobra
    Naja naja naja — Indian cobra
    Naja naja oxiana — Oxus cobra
    Ophiophagus hannah — king cobra
  Family — *Viperidae* — vipers, adders
    Vipera berus — European adder
    V. russelli — Russell's viper
    V. xanthina palaestinae — Palestine viper
    Echis carinatus — saw-scaled viper
  Family — *Crotalidae* — pit vipers
    Trimeresurus flavoviridis — Okinawa habu
    Trimeresurus spp. — Asian lance-headed vipers–30 species
    Agkistrodon halys — Mamushi
    Agkistrodon spp. — 9 species

FIGURE 1. Asian cobra, *Naja naja naja*.

FIGURE 2. Oxus cobra, *Naja naja oxiana.*

snake behavior.[1] They are oviparous, laying eggs in a nest which they remain near and may defend.[1] The generic name derives from the prey preference of the king cobra which is primarily other snakes.

2. Family *Viperidae*

Russell's viper (tic-palonga), *Vipera russelli,* (Figure 4) is the leading cause of venomous snakebite in humans in India and Burma, but the fatality rate is lower than that resulting from bites from kraits, cobras, and saw-scaled vipers. Russell's vipers are small snakes, with an average length ranging from 1.02 to 1.27 m (40 to 50 in.) to a maximum of 1.52 m (65 in.). The basic body color is deep yellow, tan, or light brown, with three rows of distinctive dark rings (Figure 4).[1] Russell's vipers are generally nocturnal and phlegmatic in disposition, but when trodden upon will strike quickly with great force.[1] Russell's vipers are found in India, eastern West Pakistan, Sri Lanka, Burma, Thailand, some of the Indonesian islands, southeastern China, and Taiwan (Figure 5).

The saw-scaled viper (carpet viper), *Echis carinatus,* has a broad distribution, from Sri Lanka, southern India, across southern Asia, the Middle East, North Africa, and south to tropical Africa (Figure 6). Russell[3] quotes Warrell[4] as saying that, "This snake probably bites more people than any other species of snake" within its range. The saw-scaled viper is a stout-bodied but short snake, only 25 to 35 cm (14 to 20 in.) long. Its body color varies considerably

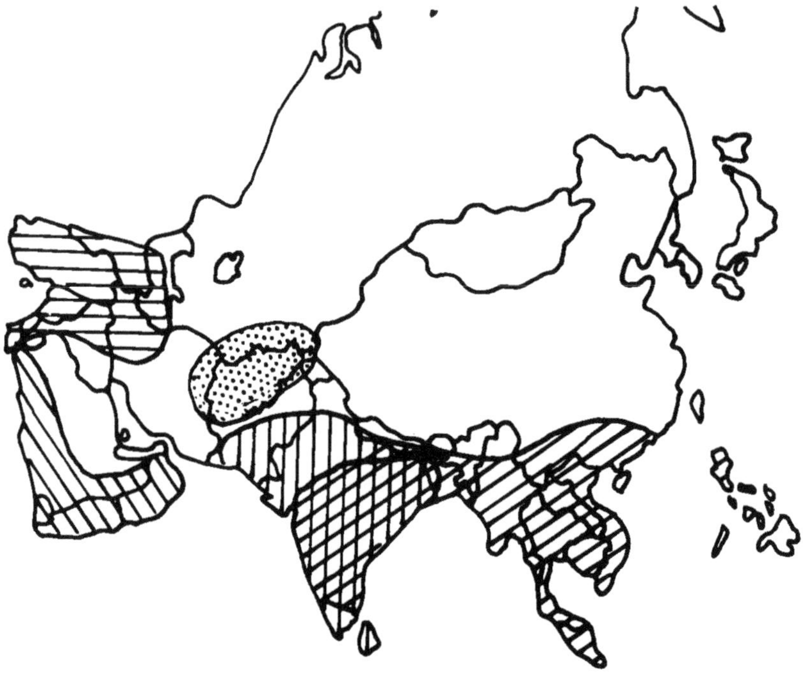

FIGURE 3. Approximate distribution of: ▤ Palestine viper, *Vipera xanthina palaestinae*. ▥ Asian cobra, *Naja naja naja*. ▩ Oxus cobra, *Naja naja oxiana*. ▨ King cobra, *Ophiophagus hannah*. ▧ Saw-scaled viper, *Echis coloratus*. (Modified from Leviton, A. E., *Venomous Animals and Their Venoms, Vol. 1. Venomous Vertebrates*, Bucherl, W., Buckley, E. E., and Deulofeu, V., Eds., Academic Press, Orlando, FL 1968. With permission.)

throughout its range, from a pale buff or tan to olive-brown, chestnut, or even reddish. Dark blotches provide variable patterns.[3]

3. Family *Crotalidae*

Asian pit vipers are represented by two genera, *Agkistrodon* (9 species) and *Trimeresurus* (30 species). Pit vipers are highly variable in size, coloration, disposition, and potency of venom. Human fatalities from bites are rare except for bites from the sharp-nosed pit viper *(Agkistrodon acutus)* and the Okinawa Habu *(Trimeresurus flavoviridis)*.

II. CONDITIONS OF ENVENOMATION

Conditions are the same as for other parts of the world. Many snakes are nocturnal and accidental encounters may occur when humans or animals are about during the night. Many species enter human and animal habitations in search of rodents.

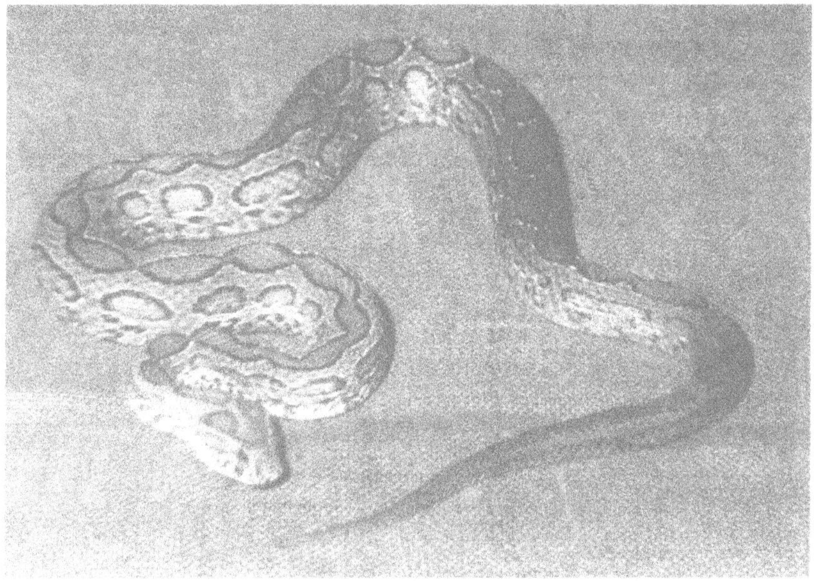

FIGURE 4. Russell's viper, *Vipera russelli*.

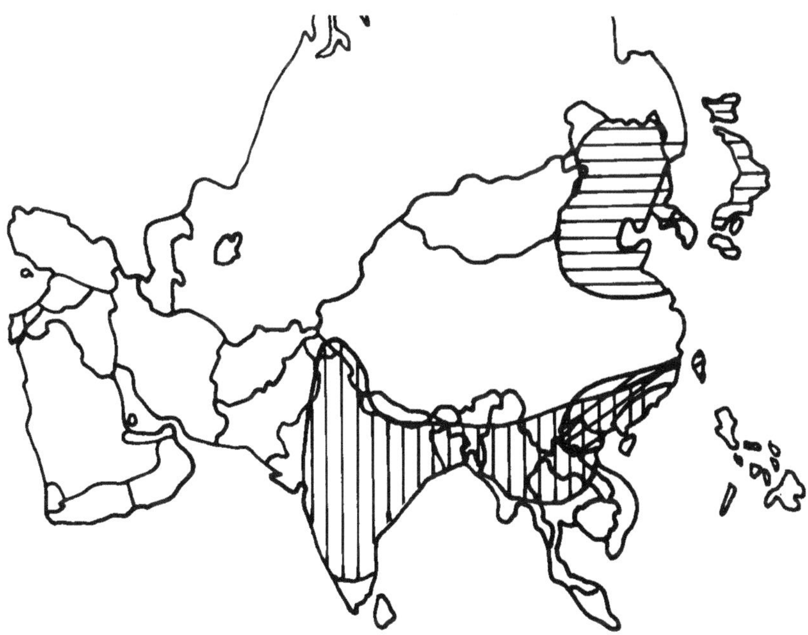

FIGURE 5. Approximate distribution of: ▥ Russell's viper, *Vipera russelli*. ▤ Mamushi, *Agkistrodon halys*. ▨ Sharp-nosed pit viper, *Agkistrodon acutus*.

FIGURE 6. Approximate distribution of saw-scaled viper, *Echis carinatus* in the Middle East and India. (Data from Leviton, A. E., *Venomous Animals and Their Venoms, Vol. 1. Venomous Vertebrates,* Bucherl, W., Buckley, E. E., and Deulofeu, V., Eds., Academic Press, Orlando, FL, 1968. With permission.)

III. VENOM[5-7]

A. ELAPID VENOM

Cobra and krait venoms are primarily neurotoxic in action.

B. VIPERID AND CROTALID VENOMS

Venoms from viperid and crotalid snakes are more variable in action than are elapid venoms, but generally are cytotoxic, necrotoxic, proteolytic, and many contain blood anticoagulant and/or coagulant factors. Table 2 lists relative toxicities of selected Asian snakes.

IV. SIGNS OF ENVENOMATION

A. HUMAN BITE[8,9]

Bites from elapid snakes are generally characterized by signs referable to neurotoxins, as described for the Egyptian cobra in Chapter 19. Local swelling and tissue necrosis may result in some cases of sufficient severity to overshadow the neurotoxic signs.[3] The victim of a single fang penetration of a finger by an Oxus cobra developed swelling of the hand and wrist in spite

TABLE 2
Relative toxicities of selected Asian snake venoms

Snake	LD_{50} mg/kg mouse, i.v.
Krait *Bungarus multicinctus*	0.07
Egyptian cobra *Naja haje*	0.42
Asian cobra *Naja naja naja*	0.04
Oxus cobra *Naja naja oxiana*	0.96
King cobra *Ophiophagus hannah*	1.60
European viper *Vipera berus*	0.55
Russell's viper *Vipera russelli*	0.30
Palestinian viper *Vipera palestinae*	0.30
Saw-scaled viper *Echis carinatus*	1.2
Asian lance-head pit viper *Trimeresurus* spp.	0.37–4.25
Mamushi *Agkistrodon halys*	7–11

of early intravenous administration of 70 ml antivenin. The victim was unable to return to work for 4 months and experienced residual stiffness for 8 months.[10]

Bites from the saw-scaled viper, *Echis carinatus*, are characterized by intense burning pain and swelling at the bite site plus hemorrhage from the gingiva, kidney, lungs, or nostril. In diagnostic examinations, physicians may ask suspected saw-scaled viper bite victims to cough deeply to determine the presence of hemoptysis.[8,9,11] Hemorrhage into vital organs such as the heart or brain may cause sudden death. The site of the fang penetration continues to exude a serosanguinous fluid. Additional clinical signs observed in human victims are listed in Table 3. Death may occur in as little as 27 h or days later from the effects of hemorrhage or secondary infection.

B. ANIMAL BITE

References to animal bites are few.[12-14] Experimental envenomation of five calves by Indian cobras, *Naja naja*, resulted in the following clinical signs: local swelling, salivation, and tremors. One calf developed hematuria. All the calves developed dyspnea and convulsions immediately prior to death, which was attributed to respiratory paralysis.[12] Similar signs were observed by Mahender et al.[13] in experimental envenomation of buffalo calves, and by Leclerc in natural cobra bite in calves.[14] The venoms of these snakes have been studied in laboratory animals and the effects appear to be similar to those observed in human victims.[15]

V. DIAGNOSIS

Venoms of vipers, particularly of the saw-scaled viper, cause coagulation defects. It is appropriate to submit blood for a clotting panel to check for prolonged clotting time, decreased platelet count, and anemia. Otherwise, the

TABLE 3
Clinical Signs of Envenomation by Selected Asian Snakes

Signs and conditions	**Russell's viper** ***Vipera russelli***	**Saw-scaled viper,** ***Echis carinatus***	**Asian lance-headed viper,** ***Trimeresurus***
Swelling and edema at site	+3	+3	+3
Pain	+3	+3	+2
Discoloration	+3	+2	+2
Vesiculation	+2	+2	+3
Ecchymosis	+2	+2	+3
Prolonged clotting time	+3	+3	+2
Hemorrhage	+2	+3	+2
Hypotension	+2	+2	+2
Nausea and/or vomiting	+2	+1	0
Necrosis	+2	+1	+1

Adapted from Moore, G. M., *Poisonous Snakes of the World,* Tables 2 and 3. NAVME P-5099, Revised 1965, Department of the Navy, Bureau of Medicine and Surgery, U.S. Government Printing Office, Washington, D.C., 1965.

clinician must rely on clinical signs described above for human envenomation and findings at necropsy on animals that die.

VI. TREATMENT[9]

Appropriate antivenin is the specific treatment of choice (Chapter 14). As respiratory failure is the common cause of death in elapid snake envenomation, it may be necessary to assist respiration.

REFERENCES

1. **Moore, G. M.,** *Poisonous Snakes of the World,* NAVMED P-5099, Revised 1965, Department of the Navy, Bureau of Medicine and Surgery, U.S. Government Printing Office, Washington, D.C., 1965.
2. **Leviton, A. E.,** The venomous terrestrial snakes of east Asia, India, Malaya and Indonesia, in *Venomous Animals and Their Venoms, Vol. 1. Venomous Vertebrates,* Bucherl, W., Buckley, E. E., and Deulofeu, V., Eds., Academic Press, Orlando, FL, 1968, 529.
3. **Russell, F. E.,** *Snake Venom Poisoning,* Scholium International, Great Neck, NY, 1983.
4. **Warrell, D. A.,** Poisoning by bites of the saw-scaled or carpet viper *Echis carinatus* in Nigeria, *Q. J. Med.,* 46, 33, 1977.

5. **Chaudhuri, D. K., Maitra, S. R., and Ghosh, B. N.,** Pharmacology and toxicology of the venoms of Asiatic snakes, in *Venomous Animals and Their Venoms, Vol. 2. Venomous Vertebrates*, Bucherl, W., Buckley, E. E., and Deulofeu, V., Eds., Academic Press, Orlando, FL, 1971, 3.
6. **Ghosh, B. N. and Chaudhuri, D. K.,** Chemistry and biochemistry of the venoms of Asiatic snakes, in *Venomous Animals and Their Venoms, Vol. 1. Venomous Vertebrates*, Bucherl, W., Buckley, E. E., and Deulofeu, V., Eds., Academic Press, Orlando, FL, 1968, 577.
7. **Khole, V.,** Toxicities of snake venoms and their components, in *Handbook of Natural Toxins, Vol. 5. Reptile Venoms and Toxins*, Tu, A. T., Ed., Marcel Dekker, New York, 1991, 405.
8. **Reid, A.,** Symptomatology, pathology and treatment of land snake bite in India and Southeast Asia, in *Venomous Animals and Their Venoms, Vol. 1. Venomous Vertebrates*, Bucherl, W., Buckley, E. E., and Deulofeu, V., Eds., Academic Press, Orlando, FL, 1968, 611.
9. **Reid, H. A. and Theakson, R. D. G.,** The management of snakebite, *Bull. World Health Org.*, 61, 885, 1983.
10. **Fowler, M. E.,** Cobra snakebite in a keeper, *Annual Proceedings*, American Association Zoo Veterinarians, 1979, 8.
11. **Deoras, P. J.,** The story of some Indian poisonous snakes, in *Venomous Animals and Their Venoms, Vol. 2. Venomous Vertebrates*, Bucherl, W., Buckley, E. E., and Deulofeu, V., Eds., Academic Press, Orlando, FL, 1971, 19.
12. **Akhtari, A. and Singh, B.,** Experimental cobra envenomation. I. Clinical haematological and biochemical changes in cross-bred calves, *Ind. Vet. J.*, 3, 21, 1983.
13. **Mahender, M., Reddy, V., Ramkrishna, K., and Reddy, E.,** Haematological studies on experimental buffalo calves injected with cobra venom *Naja naja*, *Ind. J. Animal Health*, 16, 117, 1977.
14. **Leclerc, J.,** Snake bite in Zebu calf in French Guinea, *Rec. Med. Vet.*, 146, 863, 1965.
15. **Tungthanathanich, P., Chaiyabutr, N., and Sitpija, V.,** Effect of Russell's viper *Vipera russelli siamensis* venom on renal hemodynamics in dogs, *Toxicon*, 24, 365, 1986.

Chapter 21

VENOMOUS MAMMALS

I. IDENTIFICATION

Only a few mammals have evolved venoms and an apparatus for injecting them into a prey species or an enemy (Table 1).

A. MONOTREMES

The duck-billed platypus, *Ornithorhynchus anatinus,* inhabits rivers and streams east of the Great Dividing range in Australia, along the southeastern border of Australia, and in Tasmania (Figure 1). Platypuses are solitary and live in burrows along the shores of streams.[1]

The short-beaked echidna exists as a number of subspecies and is distributed throughout Australia and Tasmania.

1. Venom Apparatus

The male duck-billed platypus has a prominent, venomous spur located on the medial aspect of the tarsus. It is slightly curved with a sharp, tapered tip and conical body.[2] The spur is used in inter-male encounters and against would-be predators. Normally, the spur is retracted against the leg; but when aroused, a muscular system causes the spur to become erect at a 90° angle to the leg. The spur is connected to a duct leading first to a reservoir nearby, then to a kidney-shaped, crural venom gland located on the medial aspect of the thigh.[2]

Echidnas have a spur, but a nonfunctional gland.

B. INSECTIVORES

A few of the hundreds of species of insectivores produce toxins in modified submaxillary salivary glands (Table 1). Ducts from the glands extend forward to the area of the incisor teeth that may be modified with channels to direct the venom into the bite wound.

II. CONDITIONS OF ENVENOMATION

Human envenomation occurs only with improper handling of these animals. Usually, the hand or arm is envenomated. Theoretically, predators of the duck-billed platypus and echidna may be envenomated, but no reports exist of such envenomation. The platypus kicks to insert the spur. Dogs have been spurred when attacking or investigating a platypus.[2]

The venom of shrews is likely to be of consequence only against other shrews or small rodents.

TABLE 1
Classification of Venomous Mammals

Phylum — *Chordata* — vertebrate animals
Class — *Mammalia* — mammals
Order — *Monotremata* — monotremes
  Family — *Ornithorhynchidae* — platypus
    Ornithorhynchus anatinus — duck-billed platypus
  Family — *Tachyglossidae* — echidnas or spiny anteaters
    Tachyglossus aculeatus — short-beaked echidna
    Zaglossus sp. — long-beaked echidna
Order — *Insectivora* — moles, shrews, hedgehogs
  Family — *Soricidae* — shrews
    Blarina brevicauda — short-tailed shrew
    Neomys fodiens — European water shrew
    Sorex cinereus — masked shrew
  Family — *Solenodontidae* — solenodons
    Solenodon paradoxus — Haitian solenodon
  Family — *Macroscelididae* — elephant shrews

FIGURE 1. Distribution of duck-billed platypus, *Ornithorhynchus anatinus*.

III. VENOM

A. PLATYPUS

Little is known about platypus venom because it possesses no unique biotoxic properties and is relatively unimportant in human envenomation. The venom does contain proteins. There appears to be seasonal variation in the toxicity of the venom that may coincide with the breeding season.[2]

B. INSECTIVORES

Shrew venom is proteolytic, with neurotoxic qualities. The composition is not precisely known; toxicologic studies have utilized whole venom. A single short-tailed shrew may have sufficient venom to kill 200 mice.[2]

IV. CLINICAL SIGNS

A. PLATYPUS

1. Human

The victim experiences immediate, intense pain and numbness around the wound. Swelling develops about the wound and progresses proximally up the arm. Axillary lymph nodes may become swollen and painful on palpation. A feeling of faintness has been described by some victims. No fatalities have been reported.[2]

2. Animal

Anecdotal reports exist of lethal envenomation of dogs; however, most authors doubt such claims.[2] Presumably, the syndrome in the dog is similar to that seen in humans.

B. INSECTIVORES

Shrews use venom to paralyze and kill prey species. The syndrome as described in experimentally injected mice includes general depression, increased urination, partial paralysis, cyanosis, dyspnea, and convulsions just prior to death.[2] A bite in a human from a short-tailed shrew has been described as producing a burning sensation and subsequent minor swelling.[3]

V. DIAGNOSIS

Diagnosis is dependent upon a history of exposure to one of the mammals.

VI. PROGNOSIS

Human or domestic animal envenomation is not fatal.

VII. PREVENTION

Avoid careless handling of known venomous mammals.

REFERENCES

1. **Strahn, R.,** *Complete Book of Australian Mammals,* Angus & Robertson, Sydney, 1983.
2. **Caras, R.,** The venomous mammals, in *Venomous Animals of the World,* Prentice-Hall, Englewood Cliffs, NJ, 1974, 317.
3. **Gibbons, W., Haynes, R. R., and Thomas, J. L.,** *Poisonous Plants and Venomous Animals of Alabama and Adjoining States,* University of Alabama Press, Tuscaloosa, AL, 1990.

SPECIES INDEX

Refer to Subject Index for common name listings.

A

B

C

U

V

Z

SUBJECT INDEX

C

N

O

P

T

For Product Safety Concerns and Information please contact our EU representative GPSR@taylorandfrancis.com
Taylor & Francis Verlag GmbH, Kaufingerstraße 24, 80331 München, Germany

www.ingramcontent.com/pod-product-compliance
Ingram Content Group UK Ltd.
Pitfield, Milton Keynes, MK11 3LW, UK
UKHW021514250926
475702UK00043B/492
* 9 7 8 1 3 1 5 8 9 8 4 3 8 *